FROM MATHEMATICS TO LOGIC IN LOGIC IN MATHEMATICS

BOOLE AND FREGE

ALIOU TALL

Docent Press

Docent Press
Boston, Massachusetts, USA
www.docentpress.com

Docent Press publishes books in the history of mathematics and computing about interesting people and intriguing ideas. The histories are told at many levels of detail and depth that can be explored at leisure by the general reader.

Cover design by Brenda Riddell, Graphic Details.

Produced with TeX. Textbody set in Garamond with titles and captions in Bernhard Modern.

© Aliou Tall 2014

All rights reserved. No part of this book may be reproduced or utilized in any form or by any means, electronic or mechanical, including photocopying and recording, or by any information storage and retrieval system, without permission in writing from the author.

Contents

1 Introduction 1
   1.1 Remarks on Gillies' 'The Fregean Revolution in Logic' 1
   1.2 Outline of the Book 9

2 Background to Boole's Logical Calculus   13
   2.1 A Résumé of Aristotelian Logic 14
   2.2 Logical Ambience in Britain Before Boole 25
   2.3 The Move Towards Abstractness in Algebra 31

3 Boole's Computational Procedure in Logic   43
   3.1 Boole's Algebraic Notation 46
   3.2 Boole's Procedures of Computation 51
   3.3 Boole's 'Primary Propositions' 65
   3.4 Systematisation and Generalisation of Aristotelian Logic 70
   3.5 Boole's 'Secondary Propositions' 80
   3.6 Some Criticisms of Boole's Logical Calculus 93

4 Some Further Developments of Boole's Logical Calculus   107
   4.1 Venn's Diagrammatic Representation 108
   4.2 Modern Algebra of Logic 121
   4.3 Axiomatic Methods for 'Boolean Algebras' 128

5 An Account of the First Two Parts of Frege's *Begriffsschrift*   137
   5.1 What is Logic About? 138
   5.2 The Project of *Begriffsschrift* 144
   5.3 A Formula-Language For Pure Thought 151
   5.4 A Formal System of Logic 178
   5.5 The Place of *Begriffsschrift* in the History of Logic 186

6	Boole–as–Frege–Discusses–Him	191
	6.1 The Standard Account of The Discussion	193
	6.2 Boole and Frege: The Same Subject–Matter	197
	6.3 The Relation of Logic to Mathematics	203
	6.4 The Nodal Point of Frege's Discussion of Boole	208
	6.5 A Picture of Boole–as–Frege–Discusses–Him	219
7	Metamathematics: A Return to Boole's Research Programme	221
	7.1 The Importance of Boole in The History of Logic	223
	7.2 Boole's Anticipation of Metamathematics	232
	7.3 Löwenheim's Revival of Boole's Research Programme	239
	7.4 Back to Gillies' 'The Fregean Revolution in Logic'	244
	7.5 An Overall Portrait of Boole and Frege	247
References		250

Chapter 1

Introduction

> *O my Lord! Advance me in knowledge.*
>
> Surah 20, Taha, verse 114

> History, if viewed as a repository for more than anecdote or chronology, could produce a decisive transformation in the image of science by which we are now possessed.
>
> Thomas Kuhn

1.1 Remarks on Gillies' 'The Fregean Revolution in Logic'

In the historiography of science concerned with the historical development of knowledge, there has been a fixation on empirical sciences for illustrations. For instance, although Kuhn himself acted as a commentator in a session in which the growth of mathematics was discussed in the light of his work at the fiftieth-anniversary meeting of the 'History of Science Society' held in 1974 at Norwalk, he never wrote anything about non-empirical sciences. Therefore, the book, *Revolutions in Mathematics* edited by Donald Gillies is more than welcome. In particular, in the paper on 'The Fregean Revolution in Logic', Gillies espouses Kuhn's general analysis of revolutions in science in order to approach the Fregean revolution.

Gillies holds that Frege (1848–1925) is a revolutionary in logic and that his revolution can be compared to 'the Copernican revolution in astronomy and physics' in which the Aristotelian-Ptolemaic paradigm was 'overthrown and irrevocably discarded' and replaced by the Newtonian paradigm. Thus Gillies appears to be reliant upon Kuhn's concept of 'paradigm', which has in his view just the right degree of precision for the analysis of revolutions in science and mathematics (Gillies 1992, p. 270).

But Gillies does not endorse two aspects of the analysis of Kuhn's concept of paradigm, namely incommensurability and the birth of a paradigm in a 'flash of intuition'. Unlike Kuhn, who denies the possibility of comparing two different paradigms, he considers that 'in a scientific revolution the old paradigm is compared with the new, and judged on perfectly rational grounds to be inferior' (Gillies 1992, p. 266). Regarding the second aspect of Kuhn's analysis, he suggests replacing the romantic theory of the birth of paradigms by a more prosaic view which he finds in the concept of 'research programme' developed in the Popperian school.

Gillies then argues that 'it is work on a research programme by a small group, or, in the limit, a single individual, that gives rise to a new paradigm' (Gillies 1992, p. 286). For him, 'a change in programme marks the beginning of a revolution if the programme leads to the development of a new paradigm' (Gillies 1992, p. 286). Thus, he claims that the programmes of Frege and Peano were revolutionary in that they generate a new paradigm, whereas Boole's research programme was not revolutionary. Gillies argues that although Boole's logical calculus does suggest extensions of traditional logic, there is nothing in the programme likely to bring about a dramatic alteration in the content of traditional logic (Gillies 1992, p. 287). He points out that forty nine percent of Boole's *Mathematical Analysis of Logic* deals with traditional logic in order to show the conservative rather than revolutionary nature of Boole's advance[1]. He says

> Peano should, like Frege, have been led by his research programme to make advances in logic going beyond anything achieved by Boole. (Gillies 1992, p. 291)

1.1.1 Gillies' General Thesis

When Gillies holds that the Fregean revolution in logic was analogous to the Copernican revolution in astronomy and physics, he refers to Kuhn's concept

---

[1]It would have been fair if Gillies had added that *The Laws of Thought* contains 424 pages and 22 chapters, and only one chapter of 9 pages is devoted specifically to Aristotelian logic.

of 'extraordinary science'. Gillies considers as three standard examples of revolutions in logic, the Aristotelian, the Fregean, and the Gödelian revolutions. He regards Aristotelian logic as the old paradigm which was replaced with the new paradigm of Fregean logic, whose dominance was then ended by the publication of Gödel's incompleteness theorems in 1931. Hence he assumes that there are only occasional revolutions in logic, and that 'normal logic' characterised by routines and puzzle-solving prevails within each of these paradigms. Thus, on his view, logicians can rest within their cocoons from which they gain an intellectual protection until one or a few of them working on a particular research programme point out 'anomalies' leading to the development of a new paradigm. This research programme is then called a 'revolutionary' research programme.

In opposition to this picture of the development of logic, I shall claim that although Kuhn's concept of 'scientific revolutions' can be considered as one of the core units for discussing the growth of logic, it does not have the importance that Gillies seems to associate with it. Indeed, regarding the history of mathematical logic, it can be suggested that logical revolutions are not so revolutionary and 'normal logic' not so normal as Gillies' analysis suggests. For a basic feature of the development of mathematical logic is that there is a sense in which knowledge gained within earlier logical research programmes is preserved in their successors. In fact, there is an overlapping of logical research programmes which makes the focus on revolutionary periods misleading. Furthermore, the discussion of the conceptual foundations of any logical research programme is a historically continuous process which makes it difficult to single out any extended historical episode when only one logical research programme stood alone. It follows that the emergence of a logical research programme does not often exhibit the unanimity of adherence or the abandonment of critical discourse which, in Kuhn's eyes, mark the aftermath of a scientific revolution.

My proposal is to use the concept of 'research programme'[2] as a tool for the analysis of the development of mathematical logic. Thus I hold that a familiar phenomenon in the history of logic is the redevelopment and extension of earlier logical research programmes. For instance, each of Boole (1815-1864), Löwenheim (1878–1940) and Hilbert (1862–1943) initiated a new research pro-

---

[2] Here I use the concept of 'research programme' in a sense akin to the Popperian school: to indicate not a dominant theory, but rather 'a mode of explanation which is considered so satisfactory by some scientists that they demand its general acceptance' (Popper 1970, p. 55). However, I shall not adhere to Lakatos' specific ideas about research programme, i.e., the concepts of 'hard core' and 'positive heuristic' (see Lakatos 1970). The use of the latter is very much favour by Gillies (see Gillies 1992, pp. 284–286 where he discusses research programme in general terms).

gramme which preserved much of the theoretical structure of its predecessor. They were not solving puzzles within a 'normal logic' i.e. an existing research programme. Yet, according to Gillies, their works did not lead to the development of a new paradigm, either. Gillies does indeed use the notion of 'research programme', but nonetheless for Gillies research programmes either lead to revolutionary advances or they take place within an old paradigm. But these are not the only alternatives and most work in mathematical logic is neither strictly revolutionary nor just 'normal science'. The image according to which all logicians spend most of their time doing what Kuhn calls 'puzzle-solving'- except in rare periods of crisis - does not seem to capture the actual development of mathematical logic.

Once it is suggested that the 'normal' state of logic is in fact characterised by the emergence of new research programmes, and the criticism and extension of older ones, the task which the historian of logic has to set himself should be the study of specific logical research programmes and the discussions of their relative merits. Hence, instead of exaggerating the difference between 'normal' and 'revolutionary' logic which probably has led Gillies to focus on 'periods of revolutionary activity', I shall discuss the development of mathematical logic from Boole to Frege in terms of overlapping research programmes.

A logical research programme is to be conceived as a set of guidelines which is accepted by some logicians for the construction of specific logical theories. It exhibits a leading principle and encompasses basic concepts, a special symbolism or terminology. In this sense, Boole's *The Mathematical Analysis of Logic* (1847) generated a research programme whose leading principle was the application of mathematics to logic. Until about 1900 this programme were pursued by A. Morgan (1806–1876), J. Venn (1864–1923) W. S. Jevons (1835–1882) and specially C. S. Peirce (1839–1914) and Ernst Schröder (1841–1902). During the latter part of the same period a new research programme emerged in Frege's *Begriffsschrift* (1879) whose leading principle was no longer the application of mathematics to logic but, conversely, the application of logic to mathematics. The development of Frege's programme appeared in the *Principia Mathematica* (1910–1913) by A. N. Whitehead (1861–1947) and B. Russell (1872–1970).

However, the basis of Frege's programme was provided by the works of the nineteenth century algebraic logicians within Boole's research programme. In effect, Boole and his followers made explicit rules of deduction by developing a purely symbolic system. This was extended by Frege who gave a final and definitive form to what formal deduction actually was. Thus Frege's research programme extended over and covered part of Boole's research programme. Hence there was an overlapping of Boole and Frege's research programmes which rendered the change of programme in this case quite complicated. And

Introduction 5

if this change is to be labelled 'revolutionary', then such a revolution should not be compared to the Russian revolution in which something is 'overthrown and irrevocably discarded' (see Gillies 1992, p. 269).

Gillies' conception of the history of logic raises the question of the appropriateness of Kuhn's concept of paradigm for analysing the development of mathematical logic. Where does Gillies draw the line between, on the one hand Aristotelian and Fregean paradigms, and on the other hand different articulations of these paradigms? Does he risk casting a shadow over the great work which was being done before and after Frege? Before Frege, Boole's logical calculus initiated the application of algebra to logic, which was of central influence on the later development of mathematical logic. And after Frege, the rich and stimulating debates concerning logicism, formalism, and intuitionism brought forth interesting developments that are based upon the critical treatment of logical problems. These leading schools of thought belonged to the background of Gödel (1906-1978), who then criticised all of them. Therefore, there were productive differences in play in logical activities before and after Frege, which should not be discounted. It then appears to be an oversimplification to reduce the whole logical enterprise into Aristotle, Frege, and Gödel.

Furthermore, how can Gillies pinpoint unequivocally the appearance of Frege's paradigm or the specific historical epoch of its emergence and effectiveness? He regards propositional and first-order predicate calculus as the core of the Fregean paradigm. But, as Putnam points out, the fact is that Peirce and his student O. H. Mitchell worked out the quantifier in the way in which it is now recognised.[3] For they used the symbols Σ for the existential quantifier, and Π for the universal quantifier, together with bound variables, in the context of a one-dimensional notation which is recognisable as a variant of modern notation, unlike Frege's counter-intuitive two-dimensional notation.[4] It is no surprise, then, that Löwenheim proved his theorem in Peirce's notation. Skolem (1887–1963) also used it in his model-theoretic approach to logic (1920–22). Even Gödel (1931) used Π for the universal quantifier. Thus it was Peirce and his student's notation that became known by the logical community. As Putnam stresses it, 'Frege's notation repelled everyone... Peirce's notation, in contrast, was a typographical variant of the notation we use to-

---

[3] In a paper entitled 'Peirce the Logician' and published in his book, *Realism with a Human Face* (1992), Putnam acknowledges that Frege is historically first to discover the quantifier. But Putnam attributes what he calls the effective discovery of the quantifier to Peirce and his student, by which he means that it is to Peirce's work rather than to Frege's that modern quantification theory can trace its origin (Putnam 1992, pp. 255–58). He even points out that Whitehead came to his knowledge of quantification through Peirce and his student. As for Russell, Putnam can not find out where he learnt about quantification (Putnam 1992, p. 258).

[4] See §??.

day' (Putnam 1992, p. 256). A point of fact is that when Frege published his *Begriffsschrift* in 1879, no one at the time grasped clearly enough what he had achieved. It was not until some decades later that the achievement began to generate the desired attention. His thought reached the awareness of others mostly through Peano, Russell, Carnap and Scholz. The fact that Frege's thought was in advance of his time is a problem neither anticipated nor treated by Gillies' account of the historiography of logic.

Although Gillies may be correct to believe that revolutions occur in logic, the conclusion he draws from it seems to be too strong and far-reaching. For instance, he claims that when the Fregean revolution in logic occurred the Aristotelian paradigm was discarded. However, as a matter of fact, it turns out that in the *Begriffsschrift* Frege reproduced the square of logical opposition as it stood in Aristotelian logic.[5] He also mentioned the syllogisms fElAptOn and fEsApO, and considered the syllogism bArbArA, which is universally valid if expressed into his logical system. Therefore, Aristotelian logic has not been completely discarded in Fregean logic. Indeed, logic involves theoretical development, but such a development is carried out slowly and by preserving principles it has once acquired.

1.1.2 Gillies on Peano's Research Programme

According to Gillies, if his general thesis is correct, then Peano's research programme was revolutionary in that it led to the treatment of Frege's (earlier) work as a new paradigm whereas Boole's research programme was not revolutionary. In defending his thesis Gillies argues that, in 'The Principles of Arithmetic' (1889), Peano (1858–1932) found an adequate means of expressing universal and existential quantifiers, which Boole obviously did not. However, as we have seen, it turns out that the quantifiers had already been used by Peirce and his student O. H. Mitchell in his publication 'On a New Algebra of Logic' (1883), and certainly their works were squarely in Boole's research programme. In effect therefore the use of the quantifiers does not seem here to be a good criterion for Gillies to evaluate the revolutionary character of Peano's research programme against Boole's. The truth is that, although Peano's work was of a somewhat different type, it belonged to Boole's research programme, of which it represented an important development. As Frege wrote to Peano, 'I see that you follow Boole, but use some other signs and try to make his logic fruitful for mathematics (Frege before 1891, p. 108).

Unlike Frege, Peano did not try to reduce mathematics to logic. On the contrary, he set up an axiomatic-deductive system in which the axioms were

---
[5] See §5.3.7.

Introduction

stated in the logical symbols that form his 'pasigraphy'. He analysed the methods of mathematics and tried to apply symbolical processes to the propositions of mathematics. It does not seem, as Gillies suggests, that the logic involved in Peano's axiomatisation of arithmetic was fully explicit. For, as Heijenoort pointed out, Peano did not have any rule that would play the role of a rule of detachment (Heijenoort 1967a, p. 84). As a result, we do not find in Peano's work a logical method in finished form. The first proof Peano gave in 'The Principles of Arithmetic' can illustrate this point. Instead of presenting a proof of the theorem $2 \in N$, he actually gave a list of formulas in which each one is very close to the next (Peano 1889, pp. 113–114). But, as Heijenoort noted, 'however close two successive formulas may be, the logician cannot pass from one to the next because of the absence of rules of inference. The proof does not get off the ground' (Heijenoort 1967a, p.84). Moreover, it was this shortcoming that Frege proposed to overcome by showing that mathematical proofs could be formalised. Thus he defined Peano's primitive arithmetical concepts in logical terms, and deduced his arithmetical axioms from logical axioms. Hence the conception of logic—and especially of its relation to mathematics—is essentially different in Peano and in Frege.

Like Boole, Peano was concerned with mathematical logic as a kind of auxiliary discipline of mathematics. There is an 'arithmetization' of logic in Peano which resembles the work carried out within Boole's research programme. Peano wanted to conceive mathematics as a purely formal procedure independent from the interpretation of the symbols involved. He realised that the notations of logic are not just a 'pasigraphy'; they are also a powerful instrument for analysing propositions. So, in the preface of the 'The Principles of Arithmetic', he acknowledged that he used the research of Boole. As he wrote, 'the notations and propositions of logic which are contained in numbers II, III, and IV, with some exceptions, represent the work of many, among them especially Boole' (Peano 1889, p. 102). Also, in his 'The Operations of Deductive Logic', he noted the equivalence of the calculus of sets and the calculus of propositions, and gave a treatment of deductive logic based on his study of the works of Boole, Schröder, Peirce and others scholars working within Boole's research programme. Moreover, as Gillies admits, Peano did share with Boole the fundamental thesis that the notion of class is the logically primitive notion. Hence even if Peano's intention was slightly different, the heuristic of his programme dovetailed with Boole's logical calculus.

Gillies also argues that Peano made advances in logic going beyond anything achieved by Boole in that he, like Frege, was led by his research programme far beyond the confines of traditional Aristotelian logic (Gillies 1992, p. 293). For, in his eyes, Peano made a remarkable discovery which showed that the laws of Aristotelian logic are invalid in special cases in which the extensions of some of the predicates are empty.

But the discovery that Gillies attributes to Peano was in practice pointed out by Boole. It must be admitted that Boole did not directly compare his own work with Aristotelian logic and make explicit the problem of existential import. But, in his *Mathematical Analysis of Logic*, he implicitly introduced the concept of null class. Indeed, in his algebraic notation zero appears as the symbol of such a class. He interpreted the universal proposition 'all Xs are Ys' as follows:

> As all the Xs which exist are found in the class Y, it is obvious that to select out of the Universe all Ys, and from these to select all Xs, is the same as to select at once from the Universe all Xs. (Boole 1847, p. 21) Hence

$$xy = x$$

or

$$x(1 - y) = 0.$$

Thus, as in modern class algebra, Boole interpreted the proposition 'All Xs are Ys' as non-existential in the sense of not implying the existence of members of the class denoted by the subject term.

Moreover, the comparison between Boole's algebra of classes and Aristotelian logic in regard to the interpretation of the existential import of propositions occurred in Venn's *Symbolic Logic* of 1881 (see subsection **4.1.1**). Also, in his paper 'On the Algebra of Logic' published in the *American Journal of Mathematics* of 1880, Peirce assumed that universal propositions do not, while particular propositions do, imply the existence of their subjects. (Peirce 1880, p. 23). Then, in a short paper 'The Aristotelian Syllogistics', he showed again that the distinction between a universal proposition and a particular proposition is that the former does not, while the latter does, imply the existence of their subjects (Peirce 1893 p. 280). In this interpretation, the traditional square of opposition must be subject to revision. Peirce did it as follows:

> **A** and **E**, All S is P, and No S is$P$, are true together when no S exists, and false together when part only of the Ss are P.**I** and **O**, some S is P, some S is not P, are true and false together under precisely the opposite conditions.
>
> **A** and **I**, Any S is P, Some S is P, are true together when there are $S's$ all of which are P, and are false together when there are $S's$ none of which are P. **E** and **O**, No S is P, and Some S is

not P, are true and false together under precisely the opposite circumstances... (Peirce 1893, p. 283)

Thus, the problem of existential import had become more explicit with Venn and Peirce, who worked within the research programme outlined by Boole.

Consequently, if what Gillies considers as the justification of the revolutionary character of Peano's research programme is correct, then he should consider Boole's research programme as revolutionary as well. For the use of the quantifiers and the rejection of the existential presupposition had been done within Boole's research programme. In truth, the importance of Peano's work lies in its being a transitional link between Boole and Frege's research programmes.

There is no doubt that Gillies' paper, 'The Fregean Revolution in Logic', encompasses valuable insights. The claim that there is a Fregean revolution in logic is certainly correct. The use of the concepts of the historiography of science, so as to substantiate such a claim is helpful. But I will not be considering the concept of 'revolution' as the basic category for the analysis of the history of logic which Gillies has assumed. Instead, I will be using the concept of research programme as a tool in the analysis of the development of mathematical logic from Boole to Frege.

1.2 Outline of the Book

I want to show that the astounding progress made in mathematical logic in the twentieth century was produced by two successive and overlapping research programmes. One is due to Boole, who created an algebra of logic, and the other, without wiping out the first, is due to Frege, who set up the logic of mathematics. The work of Boole and Frege gave rise to two research programmes which have given rise to important consequences, such as Löwenheim's theorem of 1915 and Gödel's proof of the incompleteness of arithmetic in 1931.

I shall expound and discuss here especially Boole's work, which was the first successful mathematical treatment of one part of logic. The criterion for selecting this approach depends on the following observation: we have yet to grasp the nature and the scope of Boole's work on logic since the few recent accounts of it are more or less sketchy, and fail to assess the relationship between the work of Boole and Frege. Hence, the historical and logical values of Boole's research programme, and its connection with Frege's, remain to be recognised.

In the next second part of my dissertation, I shall be concerned with Boole's introduction of mathematics in logic. In the second chapter, I give a résumé of Aristotelian logic, in order to acquaint the reader with the elementary concepts of logic, which Boole seeks to develop in a rigorous mathematical system. I also discuss the early innovations in logic made by Hamilton and de Morgan and the move towards abstractness in algebra, to which Boole himself contributed with his mathematical works. This shows that the mathematical analysis of logic is not a happy coincidence, but a direct result of the reform of traditional logic, and the mathematical work of Boole's compatriots and his own research on mathematics.

In the third chapter, I describe Boole's logical system in some detail. I include some mathematical and logical technicalities, which are unavoidable in order to apprehend Boole's computational procedure, and how mathematics can be introduced in logic. I deal with the system of signs of Boole's logical calculus, the formulation of its laws, the rules of the operation of this kind of algebra, and the setting up of his logical system. Then I show the way in which Boole performs the practical application of the algebraic method to logic. Subsequently, I discuss some criticisms levelled at Boole. I defend Boole against the accusation of mathematicism. I do not, however, try to defend his use of the symbol 'v' to express existential judgments. But I indicate the attempts made by Peirce, Schröder and Whitehead to work out a way of repairing the flaw of Boole's 'v'.

In the fourth chapter, I am concerned with further developments of Boole's logical calculus. I point out the recurrence of a set of logical problems, and exhibit the improvements which Venn brought to logic as algebra. Then, I give an account of the axiomatic methods for 'Boolean algebras' set up by Huntington and Sheffer, who drew from Boole's logical calculus a method whose principles have been cast into the form of postulate sets.

In the third part, I shall be dealing with the introduction of logic in mathematics. I introduce Frege's logical system in chapter five. I first discuss the concept of logic, in order to elucidate how it is understood by Frege and Boole, and to show the closeness of their views on the subject-matter. Then I describe the project of *Begriffsschrift* and discuss how the concept of formal proof was conceived by Leibniz. Subsequently, I give an account of the first two parts of *Begriffsschrift*, which encompass two major achievements: the device of a 'language of pure thought', and an axiomatisation of propositional logic. The chapter contains also some logical technicalities indispensable for the understanding of Frege's logical system. Finally, I discuss the extent to which *Begriffsschrift* is a revolution in the history of logic, and discuss the predominant current view of the place of *Begriffsschrift* in the history of logic.

In the sixth chapter, I shall discuss the relationship between Boole and Frege. I consider Frege's discussion of Boole's logical calculus, and show that in this discussion Frege just confined himself to comparing the two formula-languages whilst bearing in mind the differences between their respective purposes. I point out the domain common to the two formal languages and the logical lineage between Boole and Frege. I stress that although they did not have the same conception of the relation of logic to mathematics, their works show the close relationship between logic and mathematics, in such a way that the separation of the two ceases to be a sharp one. I also indicate Frege's different and more ambitious way of treating logic, and draw a picture of 'Boole-as-Frege-discusses-him'.

I conclude, in the fourth part, by showing the importance of Boole in the history of logic. I stress that Boole's research programme through the work of Peirce developed a propositional calculus and a predicate calculus of functions of one and of several variables with quantification. Then I inquire about the conditions under which metamathematics emerged. I argue that Boole should earn the right to be considered as the grandfather of metamathematics. The argument is substantiated, on the one hand, by developing the idea that Boole's formalist treatment of logic leads on to Hilbert's metamathematics, and on the other by regarding Boole's semantics as having suggested the model theoretic approach to logic that is prominent in Löwenheim's paper of 1915, which constitutes a revival of the Boolean research programme. This leads me back to Gillies' 'The Fregean Revolution in Logic'. I suggest that the emergence of the model-theoretic approach to logic should not be regarded as a part of the Fregean revolution, but a distinct research programme whose possibility required the research programme of Boole.

Chapter 2

Background to Boole's Logical Calculus

> *Without a grain of metaphysics, there cannot be science.*
>
> Cantor

Introduction

Mary Everest Boole[1] and Hackett[2] both alluded in their writings to the existence of some metaphysico-poetic roots of Boole's logical calculus. For instance, in the pamphlet *Boole's psychology*, Boole's wife revealed that in 1832, when Boole was about seventeen, it 'flashed upon' him as he was walking across a field that, besides the knowledge obtained from direct observation, man derives knowledge from some source undefinable and invisible—which she called the 'unconscious'.

Perhaps Boole's logical calculus exhibits a metaphysical mark. But this metaphysics should not be isolated in an inaccessible sky of ideas or entities without any contact with experience. It should be rather a kind of metaphysics set to work, anchored in the finitude of the world, and in the doings of nature. Viewed as such, metaphysics may then play a stimulating role by opening the way to science. Indeed, a scientific theory always emerges from a metaphysical

---

[1] M. Everest Boole, quoted in Laita (1977), p. 163.
[2] Hackett, *The Method of George Boole*, Proceeding of the Royal Irish Academy, vol. 57, Sect. A. pp. 78–86.

inspiration, instead of an accumulation of observations. As Popper puts it, 'scientific discovery is akin to explanatory story telling, to myth making and to poetic imagination' (Popper 1981, p. 87). However, such an inspiration occurs only within the various elements of the scientist's background which involve traditional beliefs, criticism, logic, imagination and empirical tests.

Accordingly, disregarding the kind of metaphysico-poetic interpretation of the genesis of Boole's logical calculus that has been advocated by Mrs Boole, I shall show, in this chapter, that there are objective circumstances which framed Boole's metaphysical inspiration to attempt the dressing of logic with mathematical clothes. Amongst these circumstances, on the one hand, I shall show the general logical ambience in Britain during the early nineteenth century, which led to the introduction of some innovations into the syllogistic doctrine. These early innovations in logic were made by Hamilton with the quantification of the predicate, and de Morgan with the notion of 'the universe of discourse'. On the other hand, I shall describe the gradual emergence of symbolical algebra in Britain to which Boole contributed with his early and late mathematical works.

2.1 A Résumé of Aristotelian Logic

Before proceeding to show the logical ambience, and to describe the gradual emergence of symbolical algebra in Britain, I shall give a résumé of Aristotelian logic, so as to acquaint the reader with the basic concepts of logic which Boole would develop in a rigorous mathematical system.

Categorical Propositions

Aristotle is concerned with the classification and interrelations of propositions. Categorical propositions have a subject term and a predicate term and contrast with hypothetical propositions which have two or more categoricals united by a connective. Categoricals join together exactly two categorical terms and asserts that some relationship holds between the classes they designate. They are subdivided according to quantity into universals and particulars or propositions which are indefinite; and according to quality into affirmatives and negatives. As Aristotle puts it,

> A premiss then is a sentence affirming or denying one thing of another. This is either universal or particular or indefinite. By universal I mean the statement that something belongs to all or none of something else; by particular that it belongs to some or not

Background to Boole's Logical Calculus

> to some or not to all; by indefinite that it does or does not belong, without any mark to show whether it is universal or particular, e.g., 'contraries are subjects of the same science', or 'pleasure is not good'. (Aristotle, *Prior Analytics* book I, 1)

A proposition is a 'judgement expressed in words' or a 'sentence indicative' which is either affirmative or negative: affirming or denying are the two contraries qualities of logical proposition. An Affirmative proposition is one whose copula is affirmative. A Negative proposition is one whose copula is negative. There is also the quantity of a proposition: if the predicate is said of the whole of the subject, the proposition is Universal; if of a part of it only, the proposition is Particular. Indefinite propositions can turn out to be universal or particular according to their analysis. For instance, 'contraries are subjects of the same science' can be taken as a universal: contraries, without exception, come always under the same science: then the subject is universal. On the other hand, the second example, 'pleasure is not good', can be taken as a particular: some pleasures are not good. The subject is here a particular. In addition, there are singular propositions such as 'Brutus was a Roman', which are considered as universals because in them we speak of the whole subject. In fact, however, Aristotle does not treat singular propositions in his logic.

A Universal affirmative proposition is called an **A**-proposition, a Particular affirmative proposition an **I**-proposition, a Universal negative proposition an **E**-proposition and a Particular negative proposition an **O**-proposition. These types of propositions **A, E, I, O** are tabulated in a square, with universal at the top, particular at the bottom, affirmative on the left, negative on the right. Aristotle begins to formulate the square of opposition in *De Interpretatione* 6–7, which contains three claims: that **A** and **O** are contradictories, that **E** and **I** are contradictories, and that **A** and **E** are contraries (17b.17–26):

> I call an affirmation and a negation contradictory opposites when what one signifies universally the other signifies not universally, e.g., every man is white—not every man is white, no man is white—some man is white. But I call the universal affirmation and the universal negation contrary opposites, e.g., every man is just—no man is just. So these cannot be true together, but their opposites may both be true with respect to the same thing, e.g., not every man is white—some man is white.

From these three claims, we can easily show that **I** and **O** are subcontraries: they cannot both be false. For suppose that **I** is false. Then its contradictory, **E**, is true. So **E**'s contrary, **A**, is false. So **A**'s contradictory, **O**, is true. This refutes the possibility that **I** and **O** are both false. Subalternation also follows.

Suppose that **A** is true. Then its contrary **E** must be false. But then the **E**'s contradictory, **I**, must be true. Thus if **A** is true, so must be **I**. Likewise subalternation can be demonstrated from **E** to **O**. In short, Aristotelian logic states the following rules: two contraries cannot be both true, but can both be false. Subcontraries cannot be false but can both be true. If one of a pair of contradictories is true, then the other is false; and vice versa. It is stated that if **A** or **E** is true, its subaltern is also true. The successors of Aristotle lay out these relations between **A**, **E**, **I**, **O** *in quadrata formula* as in Figure 2.1.

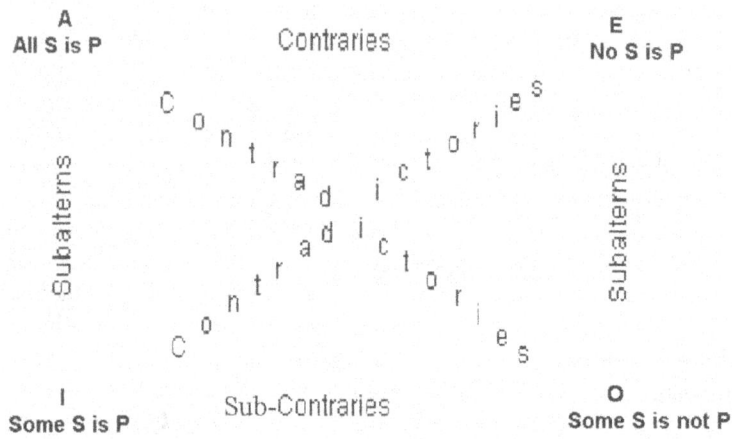

Figure 2.1: A Quadrata Formula

Later, I will show the change brought to the square of opposition by modern logicians (see subsection **4.1.1**). But why does Aristotelian's square of opposition need changing? Consider a case where S does not in fact apply to existing things. Thus the **I** form: 'Some S is P' is false. But then its contradictory **E** form: 'No S is P' must be true. But then the subaltern **O** form: 'Some S is not P' must be true. But this is quite wrong, since S is empty. In effect, Aristotelian logic differs from modern logic in that it permits the subaltern inference of 'Some S is not P' from 'No S is P' (and likewise, 'Some S is P' from 'All S is P') whereas modern logic does not permit the inference. Modern logicians hold that if there is no S, then 'All S is P' and 'No S is P' are true, while 'Some S is P' and 'Some S is not P' are false.

It is worth noting in this connection how medieval logicians approached this issue of the existential import of universal propositions. In *Truth and Consequence in Mediæval Logic*, Moody says that in practice the medieval logicians assumed the rule that all affirmative categoricals have existential

Background to Boole's Logical Calculus   17

import, as an overall postulate, without explicitly adding it to the special truth rules for sentences of the four traditional forms (E. A. Moody 1953, p. 50).

Indeed, the medieval logicians rehabilitated the traditional square of opposition through the introduction of the theory of supposition. This theory was principally concerned with what a categorical term can be taken to refer to in a particular context. Thus the four traditional forms can be said to suppose that the terms to which they refer do apply to existing things. That is the conditions of their truth or falsity are satisfied only if it is supposed that the existential question has already been answered in the affirmative. If it is supposed that all terms do apply to existing things, then the logical relations holding among them do hold.

Kneale recognises that Abelard (1079–1142), who was influential in the formation of medieval logic, had apparently accepted the ordinary account of the square of opposition. However, he holds that Abelard should have the credit of being the first to worry about the traditional square of opposition (Kneale 1962, p. 211). For he advocated the view that

> An affirmative categorical proposition was true if and if only the subject term and the predicate term both stood for the same thing or things. (Kneale 1962, p. 209)

This view was accepted by many of the successors of Abelard. Thus, in *Sophisms On Meaning and Truth*, Buridan (1300–60) stated the conditions for the truth of a particular affirmative, as well as those for the truth of a universal affirmative as follows:

> It is that every true particular affirmative is true because the subject and the predicate stand for the same thing or things. And every universal affirmative is true if whatever thing or things the subjects stands for, the predicate stands for that thing or things. (Buridan c. 1496/1500, p. 93)

Certainly, from the requirement that all the terms apply to existing things, it follows that Aristotelian's square of opposition becomes perfectly in order. As Boehner stresses, by admitting the inference 'Some man is mortal' from the proposition 'Every man is mortal', the Scholastics then insisted on the existential import of a categorical (Boehner 1952, p. 30).

Bochenski points out, as well, that the medieval logicians discussed the problem of the empty term through the notions of 'ampliation' and 'appellation'. In order to illustrate the point, he cites two thirteenth century texts, one from Peter of Spain (ca. 1210–77) and one from William of Shyreswood, and a text from Buridan as an example of the fourteenth century theories (Bochenski 1970, p. 175). It appears through these texts that medieval logicians clearly recognised and squarely faced the problem of empty terms[3]. But it seems that every time they confronted the problem they agreed with Aristotle by preserving the customary square of opposition.

Conversion and Syllogism

In order to provide universal types of reasoning, Aristotle devises two logical operations: Conversion and Syllogism.

Conversion is a type of immediate inference in which from a given proposition another proposition is drawn which has as its subject the predicate of the original proposition. In *Prior Analytics* I, 2, Aristotle writes:

> It is necessary then that in universal attribution the terms of the negative premiss should be convertible, e.g., if no pleasure is good, then no good will be pleasure; the terms of the affirmative must be convertible, not however, universally, but in part, e.g., if every pleasure is good, some good must be pleasure; the particular affirmative must convert in part (for if some pleasure is good, then some good will be pleasure); but the particular negative need not convert, for if some animal is not man, it does not follow that some man is not animal.

Thus, Aristotle observes that **E**-proposition can be converted in this way: if 'no pleasure is good', then it follows that 'no good is pleasure'. The proposition is negative and its terms are transposed: the subject of the original proposition is a predicate in the converse and its predicate a subject in the converse. **I**-proposition can be converted in the same way: if 'some pleasure is good', then it follows that 'some good is pleasure'. In later terminology such propositions were said to be converted *simply*. But **A**-proposition cannot be converted in this way; if 'every pleasure is good', it does not follow that 'every good is pleasure'. It follows, however, that 'some good is pleasure'. The proposition 'every pleasure is good' means that every pleasure is a part of good, but not every good is necessarily pleasure. We must therefore limit its quantity from

---

[3] I have chosen not to spell out the medieval logicians' attempt to solve the problem of the empty term I shall refer the reader to *A history of Formal Logic*, where Bochenski gives three other series of texts, the first attributed to St Vincent Ferrer (1350–1419), the second from Paul of Venice, the third from a neo-scholastic of the 17th century John of St Thomas. Each text gives a different solution of the problem of the empty term. For instance, Paul of Venice states the general rule that both propositions in a subalternation must have subjects with exactly the same supposition (Bochenski 1970, pp. 221–224).

Background to Boole's Logical Calculus

universal to particular; in such cases the conversion was later called by *limitation* or *per accident*. Regarding **O**-proposition, Aristotle says that it 'need not convert'; from the fact that 'some animal is not man', it does not follow that 'some man is not animal'. These laws of conversion can be used to reduce other syllogisms to syllogisms in the first figure, as described below.

Syllogism is a type of mediate inference in which from 'certain things being stated' which form the premises of reasoning, we draw necessarily something other than what is stated which will be the conclusion. In *Prior Analytics* I, 4, Aristotle gives its technical definition as follows:

> Whenever three terms are so related to one another that the last is contained in the middle as in a whole, and the middle is either contained in, or excluded from, the first as in or from a whole, the extremes must be related by a perfect syllogism. I call that term middle which is itself contained in another and contains another in itself: in position also this comes in the middle. By extremes I mean both that term which is itself contained in another and that in which another is contained. If A is predicated of all B, and B of all C, A must be predicated of all C: we have already explained what we mean by 'predicated of all'. Similarly also, if A is predicated of no B, and B of all C, it is necessary that no C will be A.

Thus, a categorical syllogism is the inference of one categorical proposition, the conclusion, from two others, the premises, each premise having one term in common with the conclusion and one term in common with the other premise. For example:

> Every animal is mortal.
> Every man is an animal.
> Therefore, every man is mortal.

The predicate of the conclusion 'mortal' is called the *major* term and the premise which contains it the major premise. The subject of the conclusion 'man' is the *minor* term, and the premise which contains it the minor premise. The term common to the two premises 'animal' is the *middle* term. In the above example that matches the relation between the three terms as described by Aristotle, we have a perfect syllogism.

But there are other types of syllogisms. Indeed, syllogisms are divided into four figures, according to the placing of the middle term in the two premises. Perfect syllogisms belong to the first figure because the middle term is subject in the major premise and predicate in the minor; in the second figure, the middle term is predicate in both; in the third figure subject in both; in the fourth predicate in the major and subject in the minor. This last figure was introduced by the successors of Aristotle who regarded the fourth figure as a redundant inversion of the first figure. The following schemata, with P for the major term, S for the minor, and M for the middle, sums up these distinctions:

1st Figure		2nd Figure		3rd Figure		4th Figure	
M	P	P	M	M	P	P	M
S	M	S	M	M	S	M	S
S	P	S	P	S	P	S	P

Within each figure, syllogisms are further divided into moods, according to the quantity and quality of the propositions they contain. Thus, in the first figure we can enumerate the following examples:

Ever X is Y
Every Z is X
Therefore, every Z is Y

No X is Y
Every Z is X
Therefore, no Z is Y

Every X is Y
Some Z is X
Therefore, some Z is Y

No X is Y
Some Z is X
Therefore, some X is not Y

These are the four valid moods of the first figure. The first has three universal affirmatives **A-A-A**; the second a universal negative, a universal affirmative and as a conclusion a universal negative **E-A-E**; the third a universal affirmative, a particular affirmative and another particular affirmative **A-I-I**; and finally the fourth mood has a universal negative, a particular affirmative and a particular negative **E-I-O**. They are called the perfect moods, and all the rest imperfect.

The Scholastics came up with a mnemonic device for representing the moods of all syllogisms commonly regarded as valid. They gave them names in which the vowels indicate the nature of the propositions that form the syllogism. Thus, the above moods are called respectively b**A**rb**A**r**A**, c**E**l**A**r**E**nt, d**A**r**II** et f**E**r**IO**. Similarly, we have the following mnemonic verses for the other three figures:

2nd Figure: cEsArE, cAmEstrEs, fEstInO, bArOcO.

3rd Figure: dArAptI, dIsAmIs, dAtIsI, fElAptOn, bOcArdO, fErIsO

4th Figure: brAmAntIp, cAmEnEs, dImArIs, fElApO, frEsIsOn.

In order to demonstrate the validity of a syllogism, we must reduce it to one of the four perfect moods. For instance, to demonstrate the validity of the syllogism cAmEstrEs of the second figure, we reduce it to the corresponding perfect syllogism cElArEnt. The letter, 'c', in cAmEstrEs, indicates that we must proceed by reducing this mood to those in the first figure beginning with the same letter. The letter, 'm', indicates that we must proceed by transfer or inversion of the premises. Finally the letter, 's', which appears twice indicates that we must proceed by a simple conversion of the proposition, which is pointed out by the vowel that precedes that 's'. Let us consider cAmEstrEs:

> Every X is Y
> No Z is Y
> Therefore, No Z is X

The inversion gives:

> No Z is Y
> Every X is Y
> Therefore, No Z is X

By simple conversion, we obtain cElArEnt:

> No Y is Z
> Every X is Y
> Therefore, No X is Z

Lukasiewicz first undertakes to rebuild Aristotle's syllogistic in a simple and rigorous way. He completely axiomatises the theory by using the modern technique of formal logic. In this way, the perfect moods of syllogism, which are self-evident, play the role of axioms of this deductive system. What is called the reduction of the other figures to those of the first, is the demonstration of these syllogisms from the axioms, considering these derivative propositions as the theorems of the system. As Lukasiewicz sees it,

> Aristotelian theory of the syllogism is an axiomatized deductive system, and the reduction of the other syllogistic moods to those of the first figure, i.e., their proof as theorems by means of the axioms, is an indispensable part of the system. (Lukasiewicz 1951, p. 44)

It follows that Lukasiewicz refutes the belief that there is a deep hiatus between Aristotelian logic and modern logic, and acknowledges it as 'a system the exactness of which surpasses even the exactness of a mathematical theory' (Lukasiewicz 1951, p. 131).

However, Lukasiewicz bases his axiomatisation upon an interpretation of Aristotelian logic which is radically different from traditional syllogism. Aristotle considers traditional syllogism as a set of propositions which are not unified so as to form a single proposition, whereas Lukasiewicz regards a syllogism as a conditional in which the premises function as a conjunctive antecedent and the conclusion as a consequent. For example, he analyses b$\mathbf{ArbArA}$ not as an inference, but as a single proposition that must be either true or false: 'If A belongs to All B and B belongs to all C, then A belongs to all C.'

This interpretation has been discredited by Lear who argues that

> The opening sentence of the *Prior Analytics* states that the scope of inquiry is proof (24a10) and one cannot make sense of the claim that a proof is a type of syllogism (25b28ff) if one treats a syllogism as a conditional. (Lear 1980, p. 9)

The argument seems to be convincing, since it is true that a proof is not a single proposition, but an inference, that is, a definite structure carried out from premisses to conclusion. Yet Lear shares with Lukasiewicz an interest that consists of exhibiting Aristotelian logic under a formal system. Indeed, he too attempts to present Aristotle's logical programme as a system of formal inference that can be subjected to mathematical examination. He even claims that when, in *Prior Analytic* A 23 and A 25 Aristotle argues that every deductive argument can be expressed as a series of syllogistic inference, he is at the same time raising the possibility of proof-theory and thus earns the right to be considered not only the father of logic, but also the grand father of metalogic (Lear 1980, p. ix).

Most of the features of Aristotelian logic which are now developed as a formal science akin in symbolism and rigour to mathematics have been introduced. What is not so far considered is the theory of forms of argument which have more than three propositions. In the book, *The Art of Thinking*, 'The Port Royal Logic' discussed the topic. Arnauld and Nicole argued that an argument can have many more than three propositions without thereby being invalid. They called such an argument a 'sorites'. As they defined it,

> A sorites is an argument composed of more than three sentences so that the first two sentences give a conclusion which when taken with the third sentence gives another conclusion, and so on. (Arnauld & Nicole 1662, p. 178)

They considered 'sorites' as the most common type of reasoning in mathematics, and relied upon the rules given for syllogisms to work it out. An example given by Stanley William (1919, p. 89) illustrates what is meant by 'sorites':

> All born within sound of Bow-bells are Cockneys
> All Cockneys are Londoners
> All Londoners are Englishmen
> All Englishmen are Teutons
> All Teutons are Aryans
> ∴ All born within sound of Bow-bells are Aryans.

In this example, it can be noticed that the predicate of one proposition is the subject of the next and the predicate of the conclusion is proved true of the subject of the first premise. This is why Cicero called the 'sorites' a 'chain of argument'. The above 'progressive sorites'[4] can be resolved into its component syllogisms. But, regarding the method of resolution, I shall refer the reader to Stanley William's *Principles of Logic* (1919, pp. 89–92) and Lewis Carroll's *Symbolic Logic & Game of Logic* (1958, pp. 84–93).

It should be stressed, however, that there is a form of argument that is not resolvable into the syllogism. Such an argument is called a 'dilemma'. A dilemma is

> A form of argument of which the major premise is composed of two hypothetical propositions; the minor premise of a disjunctive proposition; and the conclusion of either a categorical statement or a disjunctive proposition, according as the dilemma is Simple or Complex. (Stanley William 1919, p. 84)

In his *Elements of Logic* (1826, p. 90) Whateley gives the following example of the 'destructive dilemma':[5]

> If this man were wise, he would not speak irreverently of Scripture in jest;
> and if he were good, he would not do so in earnest;
> but he does it, either in jest, or earnest;
> therefore he is either not wise, or not good.

Although, as Whately showed it, every dilemma may be reduced into one or more simple conditional syllogisms (Whately 1826, p. 90), it becomes clear,

---

[4] A 'sorites' has two different forms: a progressive, which is the usual form given above, and a regressive form (see Stanley William 1919, p. 89).

[5] A dilemma has three different forms: the simple constructive dilemma, the complex constructive dilemma and the destructive dilemma (see Stanley William 1919, pp. 84–87).

with forms of argument as complicated as the dilemmas, that we are concerning with logical inferences which are more appropriately handled by formal symbolism than natural language. Hence the importance of Boole's logical calculus which makes fuller use of mathematical processes, and thereby goes substantially beyond the theory of syllogism.

The above rapid overview of Aristotelian logic shows that Aristotle developed the study of forms of reasoning by reference to some general principles which establish under an abstract form the validity of deductive inference. He carried out a research programme in logic which has had an unparalleled influence upon the history of this discipline. A prospective reading would stress some further points, which make him the precursor of modern formal logic. Amongst these points may be listed the following: the use of abstract schemas essential to formal logic; the latent idea of a formal system of logic; the general conception of logic as a formal method of science; the establishment of certain fundamental laws: the law of non-contradiction, the law of excluded middle; and the nonexistence of psychological terms in his exposition of the theory of syllogism, shunning thereby what is called 'psychologism' in logic.

Nonetheless, Aristotelian logic had a limited range. It neglected propositional logic, and so did not cover the whole field of logic. It did not explore forms of arguments, such as 'If the first then the second, the first; therefore the second', whose validity can be demonstrated by replacing the place-holders 'the first' and 'the second' by propositions. The Stoics were concerned with such arguments but did not take propositional logic to be a supplement, but rather a competitor, to Aristotelian syllogistic. In addition, Aristotelian logic did not explicitly consider the important topic of the existential import of propositions. Furthermore, because it overlooked the logical properties of the copula upon which the validity of an inference is based, it was prevented from developing a more satisfactory calculus of reasoning than the syllogism. What was especially lacking in it was a theory of relations. It was therefore inadequate for mathematical demonstrations. As de Morgan observed in his 1860 paper, 'On the syllogism: IV; and on the Logic of Relations', syllogism was inadequate for the relational reasoning that occurs often in mathematics. He wrote:

> It is not the truth that all inference can be obtained by ordinary syllogism, in which the terms of the conclusion must be terms of the premises. If any one will by such syllogism prove that because every man is an animal, therefore every head of a man is a head of an animal, I shall be ready to set him another question. (de Morgan 1850, p. 216)

Background to Boole's Logical Calculus 25

Certainly, there are forms of arguments which are employed in mathematical reasoning that cannot be resolved into the syllogism. For example, implications, such as 'If x is greater than y, and y is greater than z, then x is greater than z' do not fall into any of the types of syllogisms. As Beth stresses it, the relation was the *bête noire* of Aristotelian logic (Beth 1965, p. 47).

These limitations[6] may serve to suggest what a new research programme should include when logic is conceived more satisfactorily. Such a new research programme was outlined by Boole and de Morgan. Boole thought of logic as concerned with form of reasoning independent of the language within which it was expressed. He proposed that logical processes could only be both generalised and expedited through the systematic use of mathematical symbolism. Whilst dealing with the processes of mathematics, he realised that an indefinitely large number of valid inferences were possible that could not be evaluated by the Aristotelian methods. Boole then developed an algebra of logic which covers a totality of proofs and mode of inferences within which Aristotelian syllogistic and propositional logic are represented (though not both at once) and recognised as valid inferences. He showed that mathematical processes are applicable not only to the domain of numbers, but to the relations between classes and between propositions, and in general to any ordered domain whatsoever. Following in the footsteps of Boole, de Morgan presented in algebraic form a logic of relations. Indeed, in his 1860 paper, 'On the syllogism: IV; and on the Logic of Relations', he formalised the logic of binary relations as a generalisation of Aristotelian syllogistic and laid the foundation for the theory of relations.

2.2 Logical Ambience in Britain Before Boole

The logical ambience in Britain during the nineteenth century consisted mainly of attempts to carry out a new logic. Since traditional logic could not dovetail with the current progress of mathematics, it was widely criticised. Thus, for instance, Mill carried out a study of inductive inference to get rid of what he regarded as an inconsistency in the logical theory. In the second book of *System of Logic*, Mill considered syllogism as a process of apparent inference, and introduced induction as being 'without doubt, a process of real inference'.

---

[6] I did not mention here an important limitation of Aristotelian logic, that is the analysis of propositions in terms of subject and predicate, which is retained by Boole. This analysis conceals propositions of very different forms by considering singular propositions (i.e. 'Socrates is mortal') as 'truly universals' (i.e. 'All Greeks are mortal'). As we shall see, it is Frege's new theory of judgment which overcomes such a limitation (see subsections **5.3.2**, **5.3.3** and **6.4.1**).

In his eyes, the conclusion of syllogism does not teach us anything more than what was already included in the premises.

There were attempts to defend Aristotelian logic, such as Whateley's *Elements of Logic*, which marked a revival of logic after its discredit in the eighteenth century. For him, deductive argument is not simply a matter of word-play, a kind of verbal sword-play. He believed in the fruitfulness of syllogism, which aims 'to expand and unfold the assertions wrapt up, as it were, and implied in those with which we set out, and to bring a person to perceive and acknowledge the full force of that which he has admitted...' (Whateley 1826, p. 216). Thus, the conclusion is implicit in the premises and the utility of the syllogism is precisely to make us aware of it.

However, it was only in the second half of the nineteenth century that there were attempts to introduce some innovations in the syllogistic doctrine, so as to make it suitable for the spirit and the methods of mathematics. In opposition to the tendency to criticise syllogism, Hamilton and de Morgan claimed that the traditional forms of reasoning should not be held in contempt, but improved, and they introduced the quantification of the predicate and the idea of 'the universe of discourse'.

2.2.1 Hamilton: The Quantification of The Predicate

Hamilton was first convinced of the need to extend and correct traditional logic in 1833. In the article on 'Logic', in the *Edinburgh Review*, first published in the same year, he carried out a thorough quantification of the predicate in affirmative propositions. Then, before 1840, through his public lectures, he extended the principle of an explicitly quantified predicate equally to negatives. Finally, the 'New Analytic of Logical Forms', in which the theory was spelled out, was published in 1846.

Hamilton's expression of 'New Analytic' describes the attempt to renew the *organon* of the 'Master of Stagirite'. According to him, since the theory of the syllogism had remained where it was left by Aristotle, the 'New Analytic' intends to complete and simplify it. For, he claims:

> Aristotle, by an oversight, marvellous certainly in him, was prematurely arrested in his analysis; he began his synthesis before he had fully sifted the elements to be recomposed; and, thus, the system which, almost spontaneously, would have evolved itself into unity and order... (Hamilton 1873, p. 510)

The 'Old Analytic' had not pursued the logical analysis of language to the ultimate elements, so as to perform a systematic and rigorous method of analysis

Background to Boole's Logical Calculus   27

of propositions before rearranging them in a synthetic manner. In Hamilton's eyes, the logical analysis of Aristotle is unsatisfactory in that it does not point out the extension of their elementary forms in the analysis of propositions. Hence, in order to 'place the keystone' in the 'Aristotelic arch', Hamilton's 'New Analytic' aims to bring forth an arrangement of logical propositions, which takes into account the extension of their terms.

The principle being 'to state explicitly what is thought implicitly', it appears to Hamilton that the fundamental mistake of traditional logic is to ignore

> That the predicate has always a quantity in thought, as much as the subject; although this quantity be frequently not explicitly enounced, as unnecessary in the common employment of language; for the determining notion or predicate being always thought as at least adequate to, or coextensive with, the subject or determined notion, it is seldom necessary to express this, and language tends ever to elide what may safely be omitted. But this necessity recurs the moment that, by conversion, the predicate becomes the subject of the proposition; and to omit its formal statement is to degrade logic from the science of the necessities of thought to an idle subsidiary of the ambiguities of speech... (Hamilton 1873, p. 516)

Thus, the doctrine of the quantification of the predicate is based on the account according to which Aristotelian classification of propositions considers only the quantity of the subject, not that of the predicate. This classification distinguishes propositions into three categories, universal, particular, and singular, referring to their quantity. But, for Hamilton, such a quantitative classification is insufficient, insofar as it does not consider all the possible relations between subject and predicate. In his view, it is necessary to quantify the predicate, which is a term, so as to set up precisely all these relations.

For example, let us consider a proposition such as, 'All men are mortals'. In Aristotle's eyes, the quality of this proposition is affirmative and its quantity universal. Thus the quantity of the proposition is that of the subject 'men'. But what about the predicate 'mortals'? In order to quantify it, we employ the logical operation of conversion, and then obtain the proposition 'Some mortals are men'. When we compare the first and the second proposition, which is obtained after conversion, we shall notice that what we really wanted to say by 'All men are mortals' is 'All men are some mortals'. It follows that what was implicitly concealed in the first proposition, that is, the quantity of the predicate, is now explicitly expressed in the second.

Likewise, when we say 'All men are rational animals' what we really wanted to say is 'All men are all rational animals'. Thus the proposition, 'All men are rational animals' gives the proposition, 'All men are all rational animals' by modifying the predicate quantitatively. In the proposition, 'All men are all rational animals', there is a relation of equivalence between the extension of two terms: 'All men' and 'All rational animals'. Thus the relation between the terms constituting the proposition is extensional. It is the identity between two terms rather than the inclusion of the predicate in the subject, as it was conceived by Aristotle.

From the doctrine of the quantification of the predicate, Hamilton extends the range of classification of Aristotle's four forms:[7]

All a is all b
All a is some b
Some a is all b
Some a is some b
Any a is not any b
Any a is not some b
Some a is not any b
Some a is not some b

As a result of Hamilton's doctrine of the quantification of the predicate, the traditional canonic form of proposition, which was 'S is P', is replaced by the form '$S = P$'. Henceforth a proposition could be regarded as an equation that equates two terms according to their extensions. There follows the necessity of representing in symbolic form the precise quantitative nature of the process of predication. Thus, we could operate with propositions according to rules similar to those used in operating with elementary algebraic equations. Hence a big step forward is made toward the beginning of an algebra of logic. As Lewis puts it, 'without Hamilton, we might not have had Boole' (Lewis 1960, p. 37).

However, it seems to be a strong claim to consider Hamilton as one of the precursors of modern mathematical logic, inasmuch as he contends that the study of mathematics is at once dangerous and useless, and 'a mathematician in contingent matter is like an Owl in day-light' (Hamilton 1852, p. 652).

2.2.2 Hamilton versus de Morgan

There was a controversy between Hamilton and de Morgan about the priority in adoption of the doctrine of the quantification of the predicate. Hamilton charged de Morgan with plagiarism. I shall argue that the accusation was

---

[7]'Application of Doctrine of Quantified Predicate to Propositions', *Lectures on logic*, 1873, vol. 2, pp. 529–534.

Background to Boole's Logical Calculus

unfounded, in so far as the two theories sprang from different sources, and were independently developed. de Morgan specified this difference as follows:

> The system I now write upon does contain that extent of quantification, and though it was published before I had any knowledge even of the fact of Sir William Hamilton having a system of his own, yet I can most distinctly affirm that all my perception of complete quantification of both terms was derived *from the algebraical form of numerical quantification*.[8] (de Morgan 1850, vol. 9 p. 90)

Unlike Hamilton, who was well-known for his aversion of mathematics, which is not, for him, a road of any kind to logic, de Morgan derived his theory of the quantification of the predicate from the algebraical form of numerically definite proposition. Whereas Hamilton was confined to the idea of quantifying the predicate and the subject, de Morgan went further to introduce numerically definite propositions, which allowed him to build up numerically definite syllogisms. He exemplified the algebraical form of numerical quantification as follows:

> A numerically definite proposition is of this kind. Suppose the whole number of Xs and Ys to be known: say there are 100 Xs and 200 Ys in existence. Then an affirmative proposition of the sort in question is seen in '45 Xs (or more) are each of them one of 70 Ys': and a negative proposition in '45 Xs (or more) are no one of them to be found among 70Ys'. (de Morgan 1847, p. 164)

Thus, by using number and proportion in order to express 'some', de Morgan could specify the scope of a particular proposition, which is indeterminate between the traditional forms of reasoning. It follows that he relied upon mathematics to approach the quantification of the predicate. Therefore he was not on the same wavelength as Hamilton, whom Peirce rightly described as 'this strikingly unmathematical scholar'. The difference of sources of their two theories of quantification is evidence in favour of de Morgan who should then be discharged from Hamilton's accusation of plagiarism.

Moreover, the controversy about the priority in adoption of the doctrine of quantification was inappropriate and surprising, since the doctrine had been formulated a few years before by George Bentham in his work, *Outline of a New System of Logic* (1827). In this book, Bentham presented the eight propositions in a way very much like Hamilton's by combining the quantity and the quality of the subject with those of the predicate. He looked into the extension of the subject and the predicate, and into their different relations within the context of judgement.

---

[8] Author's emphasis

2.2.3 de Morgan: The Universe of Discourse

de Morgan's main innovation consists of the introduction of the idea of 'the universe of discourse'. His important writings on logic may be found in the *Formal logic*, first published in 1847, and a series of essays in the *Transactions of the Cambridge Philosophical Society*, amongst which is 'The Theory of Syllogism', published in 1850.

In *Formal Logic*, he introduces the notion of 'the universe of discourse' as follows:

> Let us say that the whole idea under consideration is *the universe* (meaning merely the whole of which we are considering parts) and let names which have nothing in common, but which between them contain the whole idea under consideration, be called contraries *in, or with respect to, that universe.* Thus, the universe being mankind, Briton and alien are contraries, as are soldier and civilian, male and female, &c.: the universe being animal, man and brute are contraries, &c. (de Morgan 1847, p. 42)

Thus, contrary names fill up the whole universe of discourse which varies from context to context, and thus contains only what we concur to consider at a certain time, in a certain context. The universe of discourse being animals, male and female are contraries. They are both wholly contained in that universe of discourse; no one of them fills alone the universe of discourse, or applies to everything in it. The universe of discourse is presupposed in ordinary language, when, for instance, a teacher says 'everybody is here'. He presupposes then that 'everybody' is restricted to persons registered on the course. It follows that, in order to understand properly a general proposition, it is always important to restrict ourselves to a universe of discourse which is the range of objects under consideration in a defined context.

However, if the universe of discourse is implicitly presupposed in ordinary language, it must then be explicitly definite in logic, so as to allow no dispute whatsoever about what does or does not belong to it. Thus, de Morgan specifies the notion:

> By the *universe* of a proposition, I mean the whole range of names in which it is expressed or understood that the names in the proposition are found. If there be no such expression nor understanding, then the universe of the proposition is the whole range of possible names. If, the universe being the name U, we have a right to say 'every X is Y,' then we can only extend the universe so as to make it include all possible names, by saying 'Every X which is U is

Background to Boole's Logical Calculus 31

> one of the Ys which are Us,' or something equivalent. (de Morgan 1847, p. 64)

Unlike Aristotle, who leaves out negative terms because of their indeterminateness, de Morgan introduces them into the proposition whilst presupposing a universe of discourse in which they are specified, as well as their corresponding positive terms. As he puts it,

> The introduction of contraries does, in fact, introduce a third term into the proposition; the *universe*, or *summum genus*, be it the whole universe of thought, or a conceivably separable portion of it. And it is to be particularly remembered, that every term is supposed to be *part* only of the universe: that is, to have an existing contrary in that universe. (De Morgan 1850, vol. 9 p. 91)

Thus, de Morgan considers the negative and the positive terms as embracing the whole universe. They cannot both apply at once. They are represented by large and small letters: if X stands for a positive term, x is the negative term. And everything in the universe is either X or x: and nothing is both. Thus, if X represents the term 'man', x expresses 'not-man'. The universe of discourse is a device invented so as to make the determinateness of the negative term x completely unambiguous. Only the elements that are members of X or are members of x are contained in the universe of discourse.

de Morgan's introduction of contrary terms and its corollary, the universe of discourse, allows a homogeneous presentation of logical propositions. His research with Hamilton on the extension of concepts leads happily to the possibility of representing a logical proposition by an equation, and thereby making it suitable for the methods of mathematics. Then the way is prepared for expressing the traditional forms of logic with algebraic symbols. Indeed, the outcome of Hamilton and de Morgan's innovations made possible a view of logic as being, at least in one of its parts, an algebra of classes.

2.3 The Move Towards Abstractness in Algebra

The introduction of variables, general symbols, to denote numbers by François Viète gave rise to the view of algebra as a generalisation and extension of arithmetic. In *Artem Analyticam Isagoge* published in 1591, he demonstrated the value of symbols by using plus + and minus − signs for operations, and letters to represent unknowns. He suggested using letters as symbols for quantities, both known and unknown. He employed vowels for the unknowns and consonants for known quantities. Viète's work was a great conceptual advance in that it brought forth the first systematic algebraic symbolism, which allows one to distinguish sharply between the important concept of variable and the idea of an unknown quantity.

Indeed, by contrast with the concept of number as a measure of the size of a perceivable magnitude, the concept of variable, which Greek mathematicians would regard as unthinkable, is not a measure of a variable magnitude, but a way of expressing a general relationship between magnitudes. The relation between variables expressed as an equation is called a function, which is represented by a functional equation of the form

$$y = f(x).$$

The concept of variable opened new avenues for mathematical research, which then grew in power and variety of application. As Tarski puts it,

> The invention of variables constitutes a turning point in the history of mathematics; with these symbols man acquired a tool that prepared the way for the tremendous development of the mathematical science. (Tarski 1942, pp. 13–14)

Such a tremendous development of mathematics was best exemplified by Leibniz and Newton's work. They both carried out infinitesimal methods, which led to extending the concept of algebra to deal with the formulation and properties of general axiomatic abstract systems, including arithmetical algebra, these systems being sets of elements with general operations and with a number of axioms. Thus, new algebras emerged which could describe mathematical entities that are not numbers.

2.3.1 The Emergence of Symbolical Algebra in Britain

The progressive detachment of mathematics from the world of our sense experience is exemplified by the emergence of symbolical algebra in Britain, in the first half of the nineteenth century. Many British mathematicians discussed the abstract basis of algebraic calculus. In 1830, Peacock, one of the founder of the *Analytical Society*[9], published the book *A Treatise on Algebra*; in 1833 he laid down 'the principle of the permanence of equivalent forms'. In 1840, Gregory, the editor of the *Cambridge Mathematical Journal*,[10] a personal friend

---

[9]The *Analytical Society* was formed in 1812 in Cambridge by Peacock, Babbage and Herschel who aimed to promote analytical methods in mathematics as used on the Continent that is, the notation $\frac{dx}{dt}$ of Leibniz which is opposed to Newton's fluxion notation. In 1819, the Analytical Society was replaced by the *Cambridge Philosophical Society*.

[10]The journal was founded in 1837, at the beginning, most of the contributors were from Cambridge. I may cite Ellis, de Morgan, Thomson, Cayley, Sylvester and Stokes as the most

of Boole, published a paper 'On The Real Nature of Symbolical Algebra'; in 1841 he wrote his book, *Examples of The Processes of The Differential and Integral Calculus*, which laid down the fundamental principle of the method of symbolical algebra. Almost in the same period, de Morgan wrote some essays on 'The Foundations of Algebra', and in 1849 the 'Trigonometry and Double Algebra'.

These great mathematicians came up with the discovery that the laws which govern arithmetical algebra specify a certain domain, but algebra can be understood in a more general sense, so that its calculi may be applied, often by restricting some of its particular laws, to entities that are not numbers. Thus, they gave rise to a new kind of algebra, named by Peacock 'Symbolical Algebra', and later known as Abstract Algebra. The most important principle to emerge was that the laws that govern the combination and manipulation of mathematical entities are more determinate than the possible interpretations of these entities. Such an algebra of entities, which are not numbers in any ordinary sense, changed definitively the understanding of the symbolical nature of mathematics.

Since the problem of notation played a central role in the emergence of symbolical algebra, it seems apposite to point out here the difference between the notations used by Leibniz and Newton to represent their infinitesimal methods, which led to extending the concept of algebra. In the book, *George Boole: his Life and Work*, MacHale describes the difference of the two approaches as follows:

Newton was at heart an applied mathematician for whom mathematics was a very powerful tool by means of which he hoped to construct a model of the physical universe and thereby understand its structure. It was inevitable that he should be a geometer and that his basic approach to mathematics should be intuitive and geometric. Leibniz, on the other hand, with his strong background in philosophy, was more a pure mathematician though, of course, the distinction between the disciplines was virtually non-existent in those days. He viewed the calculus as an exciting development in its own right rather than as a tool for physics. His approach to the subject was akin to that of a present-day algebraist and his methods were abstract and analytical (MacHale 1985, p. 44).

Thus, MacHale clearly favours Leibniz's approach. But what was really fundamental in their difference is the notation used to express their infinitesi-

---

famous names which were mentioned in its first series. The release of the journal coincided with the changes in British mathematics carried out by the innovators of *Analytical Society*, who needed a journal to publish their research material.

mal methods. According to MacHale, the use of the symbol

$$\frac{dx}{dt}$$

indicates that Leibniz envisaged differentiation as a process of an operator applied to the variable x; it may yield computation when the symbol

$$\frac{d}{dt} = D$$

is regarded as a mathematical entity in its own right. On the other hand, Newton's 'method of fluxions', which was more dependent on diagrams and geometric models, demonstrates that he considered the subject as physical motions of various kinds. His dot notation did not specify change in x with respect to time. So Leibniz's notation was more general.

By preferring Leibniz's notation rather than Newton's one, MacHale follows the way opened by the innovators of *Analytical Society,* namely, the algebraist George Peacock (1791–1858), the astronomer John Herschel (1792–1871) and Charles Babbage (1791/2–1871). These mathematicians, following the example of Woodhouse,[11] had shown their devotion to science beyond all chauvinism by adopting Leibniz's notation. Thus, it has been told that Babbage described the aim of the society as to promote the principles of pure 'd'ism' as opposed to the 'dot-age' of Cambridge, which preferred the dot notation of Newton. They changed the mainstream of British mathematics through the publication of their textbooks and the translations of the Continental works. To this end, they were backed up by the *Cambridge Mathematical Journal,* which published their research.

Peacock's *Treatise of algebra,* 'written with the view of conferring upon Algebra the character of a demonstrative science', was amongst these textbooks. In order to attain this aim Peacock suggested a review of the relationship between arithmetic and algebra. Instead of conceiving arithmetic as the foundation of algebra, he viewed it as 'a science of suggestion, to which the principles and operations of algebra are adapted, but by which they are neither limited nor determined.' Accordingly, he distinguished 'arithmetical' from 'symbolical' algebra. Arithmetical algebra is concerned with numbers, and the operations are those of arithmetic. As for symbolical algebra, it is 'a science, which regards the combinations of signs and symbols only according

---

[11] In 1803, Woodhouse, a Cambridge mathematician, released his *Principles of Analytical Calculation* in which he gave an account of the Continental usage of the differential notation, and required more consideration of the subject. The book happily influenced the founder of the *Analytical Society.*

Background to Boole's Logical Calculus

to determinate laws, which are altogether independent of the specific values of the symbols themselves.'

Nonetheless, Peacock worked out a way to link the two by the 'principle of the permanence of equivalent forms', which states that:

> Whatever form is algebraically equivalent to another when expressed in general symbols, must continue to be equivalent whatever these symbols denote.

Conversely,

> Whatever equivalent form is discoverable in arithmetical algebra considered as the science of suggestion when the symbols are general in their form, though specific in their value, will continue to be an equivalent form when the symbols are general in their nature as well as in their form. (Carl Boyer 1989, p. 576)

Arithmetical algebra is indeed also a collection of basic rules which form an axiomatic structure. Similarly symbolical algebra deals with the formulation and properties of general axiomatic abstract systems of this form. These systems are sets of elements with general operations and with a number of axioms.

However, symbolical algebra is based on axioms which differ from those of arithmetical algebra. Peacock had seen the necessity of turning symbolical algebra into a structure of postulate sets, which are not the same as those of arithmetical algebra. Thus the procedure of symbolical algebra does not depend upon the meaning of the symbols, but merely upon the laws of their combinations.

Following Peacock other British mathematicians went on further to envisage algebra in an abstract fashion. For instance, in the 'Trigonometry and Double Algebra', de Morgan claimed that it may be possible to build up an algebraic system with arbitrary symbols and a set of laws under which these symbols would be handled. As for Gregory he wrote an essay 'On the Real Nature of Symbolical Algebra' and published the book, *Examples of The Processes of The Differential and Integral Calculus*. In the latter, he said:

> In this chapter 'On general theorems in the differential calculus' I shall collect those theorems in the differential calculus which, depending only on the laws of combination of the symbols of differentiation, and not on the functions which are operated by these symbols, may be proved by the method of separation of symbols: but as the principles of this methods have not as yet found a place in the elementary works of calculus, I shall first state briefly the theory on which it is founded.

Then he pointed out the fundamental principle of the method as follows:

> The laws with which we have here concern are few in number, and may be stated in the following manner. Let a, b represent two operations, u, v two subjects on which they operate, then the laws are:
>
> $$ab(u) = ba(u)$$
> $$a(u+v) = a(u) + a(v)$$
> $$a^m \cdot a^n \cdot u = a^{m+n} \cdot u$$
>
> ...The third law is not so much a law of combination of the operation denoted by a but rather of the operation performed on a which is indicated by the index affixed to a... these are the laws employed in the demonstration of the principal theorems in Algebra... but they are not confined to symbols of numbers: they apply also to the symbol used to denote differentiation. (Laita 1977, 34, p.172)

Thus, Gregory laid down the method which consists of extending the ordinary laws of arithmetical algebra to symbols standing for mathematical entities that are not numbers. As a result, the principle of an extension to logical symbols was set in motion. Indeed, in *The Mathematical Analysis of Logic* (Boole 1847, pp. 15–18), Boole conceived logic as an algebra with the same distributive and commutative laws as in symbolical algebra, and the index law, waiting for interpretation. It follows that the first principles of Boole's logical system are derived from Gregory's principle of the method of symbolical algebra, when interpreted in logic.

Hence, it can be said that the ground for Boole's logical calculus was prepared by the move of the British mathematicians towards abstractness in algebra, which allowed logic to be regarded as a domain over which universal mathematical calculus is performed which embodies not only arithmetical algebra, but also other possible calculi.

2.3.2 Boole's Contributions to Symbolical Algebra

Boole wrote several papers and was one of the first to look into the basic properties of numbers, such as the commutative, the associative and the distributive properties, which underlie the discipline of algebra. Thus, he was aware of an abstract structure in algebra without necessarily any interpretation to numbers or anything else. He published a paper on 'Exposition of a General Theory of

Background to Boole's Logical Calculus

Linear Transformations' (1843) in which he gave rise to the algebraic theory of invariants. In the book *Men of Mathematics* (1953, p. 483), Bell even holds that Boole discovered invariants.[12]

I shall point out the contributions made by Boole himself to the development of symbolical algebra, so as to show that the mathematical analysis of logic is not a happy coincidence, but a direct result of the mathematical work of Boole's compatriots and his own research on mathematics.

Boole's outstanding contributions to the move towards abstractness in algebra can be found in his early and late works. In the book *George Boole: His Life and Work*, MacHale recapitulates Boole's early contributions to mathematics as follows:

> Boole contributed some twenty-four mathematical papers to the *Cambridge Journal*, twelve to each series. These papers covered a wide range of mathematical topics including differential equations, integration, logic, probability, geometry and linear transformations. However, the importance of these contributions lies not so much in their content, but in the fact that they stressed the importance of the manipulation of symbolic operators in various areas of mathematics (MacHale 1985, p. 52).

Amidst Boole's early publications, I shall single out two papers, in which he followed the mainstream of generalisation from symbolization whilst showing his own initiative by separating the symbols of mathematical operations from the subjects upon which they perform and to investigate these operations on their own account. Indeed, these papers deserve consideration here in that they exemplify what MacHale considers as the nodal point of Boole's contributions, that is, 'the importance of the manipulation of symbolic operators'.

The first paper is entitled 'On The integration of Linear Differential Equations with Constant Coefficients'. It was published in 1841 in the *Cambridge Journal*. In it, Boole gave a simplification of the so-called 'method of separation of symbols' which Gregory had applied to the *Integration of Linear Differential Equations with constant coefficients*. He exhibited a process which showed that 'the form of the solution depends solely on the method of decomposing the original operating factor; and this decomposition is effected by means of processes which are common to the two operations under consideration, being founded only on the common laws of the combinations of the symbols (Boole

---

[12]Originally, Arthur Cayley, a personal friend of Boole was considered as the inventor of invariant theory in his memoir: 'On the Theory of Linear Transformation', which was published in 1845. But, in the paper, Cayley acknowledged that: 'The following investigations were suggested to me by a very elegant paper on the same subject, published in the journal by Mr. Boole' (MacHale 1985, p. 56).

1841c, p. 119). The method according to which, once the symbols and rules are specified, one may proceed to carry out the results without interpreting every step of the process is apparent in his logical development of functions. The philosophy which was developed on the basis of this method was to 'separate' symbols from their senses and to perform operation upon them according to definite algebraic rules. This idea of formal procedure in mathematics in which the symbols and rules of operations in a particular discipline are freed from all meaning guided the setting up of Boole's logical system.

The second paper is 'On a General Method in Analysis': it appeared in *Philosophical Transactions of The Royal Society* in 1844. The paper was the most important in all Boole's publications. He was awarded a medal by the Royal Society for his contributions to analysis. In the paper, Boole extended the ordinary laws of analysis to symbols expressing mathematical entities that are not numbers. He introduced it by appreciating what had been already achieved in analysis by mathematicians such as Gregory, Servois, R. Murphy, de Morgan, &c., who paid much attention to 'the calculus of operations', or the method of 'the separation of symbols'. Then, he went on to say:

> Mr. Gregory lays down the fundamental principle of the method in these words: "There are number of theorems in ordinary algebra, which, though apparently proved to be true only for symbols representing numbers, admit of a much more extended application. Such theorems depend only on the laws of combination to which the symbols are subject, and are therefore true for all symbols, whatever their nature may be, which are subject to the same laws of combination." The laws of combination which have hitherto been recognised are the following, π and ζ being symbols of operation, u and v subjects.
>
> 1. The commutative law, whose expression is $\pi \zeta u = \zeta \pi u$.
> 2. The distributive law, $\pi(u+v) = \pi u + \pi v$.
> 3. The index law, $\pi^m \pi^n u = \pi^{m+n} u$.
>
> Perhaps it might be worth while to consider whether the third law does not rather express a necessity of notation, arising from the use of general indices, than any property of the symbol π. The above laws are obviously satisfied when π and ζ are symbols of quantity. They are also satisfied when π and ζ represent such symbols as $\frac{d}{dx}$, $\triangle$, &c., in combination with each other, or with constant quantities. (Boole 1844, p. 225)

However, Boole noted that the above method was of necessity limited in its application to only linear equations with constant coefficients. He pointed out that the 'calculus of operations' tended rather to simplify the processes of analysis than to extend its power. Hence, in the paper, he set out to 'develop a method in analysis, which, while it operates with symbols apart from their subjects, and may thus be considered as a branch of the calculus of operations, is nevertheless free from the restrictions to which we have alluded' (Boole 1844, p. 226). Thus, Boole virtually launched the possibility of an extension of the ordinary laws of analysis to logic. As MacHale sees it,

> The importance of Boole's paper of 1844 lies not so much in the results he proved (though undoubtedly they constituted quite a significant contribution to mathematics) but rather in the influence it had on his subsequent ideas and development. He was now on the threshold of his greatest discovery—namely, that the essence of mathematics consists in the study of form and structure rather than content, and that 'pure mathematics' is concerned with the laws of combination of 'operator' in their widest sense. This paper gave Boole's confidence in his own understanding of the true nature of mathematics an enormous boost and, as time went on, he became firmly convinced that he had a mission to explain to the world the nature of logic, thought and, ultimately, the workings of the human mind. (MacHale 1985, pp. 65–66)

Boole's late contributions to symbolical algebra are displayed in his books, *A Treatise on Differential Equations* (1859), and *A Treatise on the Calculus of Finite Differences* (1860). In these books, Boole worked on differential equations, and on the calculus of finite differences.[13] He developed symbolical processes free from any exclusive interpretation, and investigated whether these algebraic methods could be applied to the solution of differential and difference-equations. Boole found that they were.

---

[13] As Boole put it, 'the Calculus of Finite Differences may be strictly defined as the science which is occupied about the ratios of the simultaneous increments of quantities mutually dependent. The Differential Calculus is occupied about the limits to which such ratios approach as the increments are indefinitely diminished (Boole 1860a, p. 1). Thus, a difference equation is an equation involving the differences between successive values of a function of an integer variable. It can be regarded as the discrete version of a differential equation, which is an equation involving the first or higher derivatives of the function to be solved for. If the equation only involves first derivatives it is called an equation of order one, and so on. Equations of degree one are called linear. Equations in only one variable are called ordinary differential equations to distinguish them from partial differential equations, which are equations involving derivatives with respect to more than one variable.

Thus, in his *Treatise on Differential Equations*, he gave an account of the state of knowledge on the subject of differential equations in the second half of the nineteenth century, and thereby reiterated the leadership of British mathematicians in algorithm analysis. The fundamental principle of the differential calculus is in fact an extension of arithmetical algebra which depends upon the introduction of a new idea—the idea of limit, which Boole defined as 'a fixed value toward which some varying value may be made to approach as nearly as we please, but which it cannot be made to reach' (Boole 1849, p. 40). According to him, given the expressions of two quantities one of which X increases uniformly and the other increases with it but not necessary in a uniform manner, the differential calculus solves the general problem of expressing the limit of the ratio by working out its exact value.

Boole showed that there exist large and very important classes of differential equations the solution of which depends on some process of successive reduction. He stressed that the reduction could be effected with greatest generality by symbolical methods. Hence in Chapter XVI he attempted 'to found the methods of solution of differential equations upon the study of the modes of their formation' (Boole 1859, p. viii). He introduced the algorithm of differential operators in order to solve linear differential equations. Boyer gives a simple example to illustrate Boole's symbolical method: for working out the differential equation

$$ay'' + by' + cy = 0,$$

the equation is written in the notation

$$(aD^2 + bD + c)\,y = 0.$$

Then, regarding D as an unknown quantity rather than an operator, we solve the algebraic quadratic equation

$$aD^2 + bD + c = 0.$$

If the roots of the algebraic equation are p and q, then e^{px} and e^{qx} are solutions of the differential equation and

$$Ae^{px} + Be^{qx}$$

is a general solution of the differential equation (Boyer 1989, p. 635). Boyer then notices that 'there are many situations in which Boole, in his *Treatise on Differential Equations*, pointed out parallels between the properties of the differential operator (and its inverse) and the rules of algebra' (Boyer 1989,

Background to Boole's Logical Calculus

pp. 635–636). As Boole saw it, 'in thus expressing an operation by a symbol, in studying the laws of that symbol, and in founding processes and methods upon these laws, we introduce no strange or novel principle of Language; for it is the very office of Language to express by symbols the procedure of Thought' (Boole 1859, p. 381). Further, he said that

> In any system in which thought is expressed by symbols, the laws of combination of the symbols are determined from the study of the corresponding operations in thought. But it may be that the latter are subject to *conditions of possibility* as well as to laws *when possible*. And thus it may be that two systems of symbols, differing in interpretation, may agree as their formal laws whenever they both express operations possible in thought, while at the same time there may exist combinations which really represent thought in the one but not in the other. (Boole 1859, p. 398)

Thus, Boole drew as a conclusion the general principle that,

> The mere processes of symbolical reasoning are independent of the conditions of their interpretation. (Boole 1859, p. 399)

He noticed that, whether it be taken as belonging to the realm of *a priori* truth, or as a generalisation from experience, it would be an error to consider the principle as a mathematical principle. He claimed its place among 'the general relations of Thought and Language' (Boole 1859, p. 399).

Regarding the solution of difference equations, Boole followed the same procedure in his *Treatise on the Calculus of Finite Differences*. In chapter XIII of this book, he wrote,

> The symbolical methods for the solution of differential equations whether in finite terms or in series are equally applicable to the solution of difference-equations. Both classes of equations admit of the same symbolical forms, the elementary symbols combining according to the same ultimate laws. And thus the only remaining difference is one of interpretation, and of processes founded upon interpretation. (Boole 1860a, p. 236)

Thus, in his *Treatise on the Calculus of Finite Differences*, which was a textbook in Cambridge until the 1920s, Boole stressed the connections between difference equations and differential equations, and pointed out the power of abstract operator methods when applied to a new area of mathematics. Of particular interest for MacHale is the chapter on linear difference equations with variable coefficients, where Boole developed symbolical processes for their solution whilst using the method of 'the separation of symbols'. Then, he concludes that 'once again, Boole anticipated the trends and need of the twentieth century, because modern computers and calculating machines are based on the discrete difference equation rather than the continuous differential equation (MacHale 1985, p. 220).

It follows that the main philosophical outcome of Boole's contributions to the move towards abstractness is the productive insight that an algebraic method proceeding by purely symbolical forms can be applied to different areas of mathematics and even to logic, and then can be conceived as an abstract calculus capable of various interpretations. This calculus treats of the combinations of arbitrary symbols upon which operations are performed by means of arbitrary laws, and its validity does not depend upon the interpretation of the symbols. As we would now put it, it is a method of 'disinterpretation'.

Now how did Boole carry out the extension of symbolical methods to logic and achieve what MacHale perceives as a 'mission to explain to the world the nature of logic, thought and, ultimately, the workings of the human mind'?

Chapter 3

Boole's Computational Procedure in Logic

> *Logic is an old subject, and since 1847 it has been a hard one.*
>
> George Boolos

Introduction

Early in the Spring of 1847, the controversy between de Morgan and Hamilton, prompted Boole to conceive logic as a deep system of relations based upon facts that are not quantitative. In the preface of the *Mathematical Analysis of Logic*, Boole says:

> In presenting this Work to public notice, I deem it not irrelevant to observe that speculations similar to those which it records have, at different periods, occupied my thoughts. In the Spring of the present year, my attention was directed to the question then moved between Sir W. Hamilton and Professor de Morgan; and I was induced by the interest which it inspired, to resume the almost-forgotten thread of former inquiries. It appeared to me that, although Logic might be viewed with reference to the idea of quantity, it had also another and deeper system of relations. If it was lawful to regard it from *without*, as connecting itself through the medium of Number with the intuitions of Space and Time, it was lawful also

to regard it from *within*, as based upon facts of another order which have their abode in the constitution of the Mind. (Boole 1847, p. 1)

The outcome of such a view of logic, and of the investigations which it suggested, is displayed in Boole's pamphlet, which presents a logic based on mathematics. Thus, Boole sided with de Morgan in the controversy between Hamilton and de Morgan, concerning the origin of the theory of quantification of the predicate. For de Morgan did not confine the idea, as Hamilton did, to quantifying the predicate and the subject, but rather went further to carry out a more mathematical theory of logic. Hence, Boole, as a mathematician, found himself in favour of de Morgan's promising approach. So, whilst thinking to publish *The Mathematical Analysis of Logic*, he wrote to de Morgan, who was at the same time finishing his *Formal Logic*. Then, so as to avoid being charged again with plagiarism, de Morgan suggested to Boole that they should both publish their work without checking over their results. It has been said that the two books appeared in the bookshop on the very same day.

Although Boole's work on mathematical logic was inspired by the controversy between Hamilton and de Morgan, there is no doubt that the new interest of British mathematicians in symbolic manipulation, which opened new ways in symbolic algebra, was a more significant influence upon him. As has been said, his compatriots, such as Gregory, prepared the way for Boole to come up with symbolical methods in analysis, which operate with symbols apart from their meanings and may thus be regarded as a branch of the 'method of separation of symbols'. The method consists of working out differential equations by use of the operator D, and Boole extended its laws to symbols expressing mathematical entities that are not numerical. As a result, he built up an algebra of logic by means of what he called the 'process of analysis', that is, the process by which combinations of interpretable symbols are carried out according to well-determined rules of combination. In a postscript to 'On a General Method in Analysis', Boole showed clearly that he was aware of the existence of a universal calculus of symbols, which may be interpreted in different calculi:

> Fearful of extending this paper beyond its due limits, I have abstained from introducing any researches not essential to the development of that general method in analysis which it was proposed to exhibit. It may however be remarked that the principles on which the method is founded have a much wider range. They may be applied to the solution of functional equations, to the theory of expansions, and, to a certain extend, to the integration of non-linear

> differential equations. The position which I am most anxious to establish is that any great advance in the higher analysis must be sought for in an increased attention to the laws of the combination of symbols. The value of this principle can scarcely be overrated: And I can only regret that in the absence of books, and circumstances unfavourable for mathematical investigation, I have not been able to do that justice to it in this essay which its importance demands. (Boole 1844, p. 282)

Boole made here with anxiety the suggestion that it is possible to carry out a new calculus from his general method in analysis. He knew that a way was opened now for mathematicians who ought to be aware of its scope. Certainly, he was anxious because he understood how promising were the newly opened avenues. However, regrettably, he did not have the material essential to their exploration. At this time, Boole was on the verge of 'discovering' his algebra of logic.

Later, the logical calculus, which he drew from the general method of analysis, is presented in the introduction of *The Mathematical Analysis of Logic* as follows:

> They who are acquainted with the present state of the theory of Symbolical Algebra, are aware that the validity of the processes of analysis does not depend upon the interpretation of the symbols which are employed, but solely upon the laws of their combination... It is upon the foundation of this general principle, that I purpose to establish the calculus of Logic, and that I claim for it a place among the acknowledged forms of Mathematical Analysis, regardless that in its objects and in its instruments it must at present stand alone. (Boole 1847, pp. 3–4)

Accordingly, since the abstract grounds of symbolical algebra had been laid down by British mathematicians including Boole himself, it was possible to regard algebra as a general methodology of calculus of symbols which could be interpreted in logic. Hence Boole claimed that the ultimate forms and processes of logic, as a science, are mathematical.

Consequently, he made an epistemological shift, in that logic henceforth has nothing to do with philosophy. As he put it,

> We ought no longer to associate Logic and Metaphysics, but Logic and Mathematics. ... Logic rests like Geometry upon axiomatic truths, and its theorems are constructed upon that general doctrine of symbols, which constitutes the foundation of the recognised Analysis. (Boole 1847, p. 13)

It should be mentioned that, in the seventeenth century, Leibniz had already hinted at the idea of introducing mathematics in logic. He even believed that 'if anyone wants to write like a mathematician in metaphysics or moral philosophy there is nothing to prevent him from rigorously doing...' (Leibniz 1996, p. 261). For, according to him, there is a formal analogy between the disjunction and conjunction of concepts on the one hand and addition and multiplication of numbers on the other. Thus, primitives concepts can be expressed by prime numbers and complex concepts by non-prime numbers. The rules which govern the combination of concepts state that a term composed of simple terms can be expressed by the product of the prime numbers which stand for its simple terms. For example, in the proposition 'man is a rational animal', if 'man' is represented by m, 'animal' by a, and 'rational' by r, then the proposition is formulated as follows:

$$m = ar.$$

If a and r stand for numbers, for example $a = 2$, $r = 3$, then $m = 6$, and the above formula has as a numerical expression:

$$6 = 2 \times 3$$

This led Leibniz to envision a logical algebra which he called *Calculus ratiocinator* (see subsection **5.2.2**). However, he did not formulate the analogy between combination of concepts and multiplication of numbers precisely, and then use it as the basis of a calculus of logic.

But, what was only a suggestion by Leibniz became a reality to Boole, who embodied, in *The Mathematical Analysis of Logic* (1847) and in *The Laws of Thought* (1854), a system of logic based on the model of algebra. In this chapter, I shall expound and discuss Boole's important research programme whilst describing his computational procedure, and showing how logic can be cast in the form of a computation.

3.1 Boole's Algebraic Notation

Boole models his logical calculus closely upon arithmetical algebra, from which it differs only by the 'index law'

$$x^2 = x,$$

which is particular to the type of algebra he is presenting. Thus, Boole employs as a notation the symbols of arithmetic, and sets up a way of expressing logical relations by means of equations. His logical calculus is based upon the relation of extension and directs attention to the symbols of the expressions rather than to their contents. The symbols stand for classes, including the 'universe of discourse' and the empty class which are represented by 1 and 0 respectively.

Boole relies upon a method that depends upon three fundamental ideas: the conception of the symbols, the laws of logic formulated as rules for operations upon these symbols, and the consideration that these rules of operation are analogous to those of an algebra of the numbers 1 and 0. He claims:

> All operations of language, as an instrument of reasoning, may be conducted by a system of signs composed of the following elements,
> 1st. Literal symbols, as x, y, etc., representing things as subjects of our conceptions.
> 2nd. Signs of operations, as $+$, $-$, $\times$, standing for those operations of mind by which the conceptions of things are combined or resolved so as to form new conceptions involving the same elements.
> 3rd. The sign of identity, $=$. (Boole 1854, p. 27)

3.1.1 The System of Signs

Boole represents the elements of language as elements of an instrument of deduction by a system of signs. In it, the symbols 'x', 'y', etc., replace names, proper names or common names, adjectives and descriptive expressions. The signs for operations like '$+$', '$-$', '$\times$', are the logical operations by which we assemble parts into a whole or separate a whole into its parts. They allow us to form new concepts from given concepts.

The sign '$+$' is used for the union of two classes (though only of classes that are disjoint). For example, if 'x' refers to the set of women and 'y' to the set of men, then '$x + y$' express the set of women and men; and '$x + y = z$', is true if and only if all people(z) are (either) women(x) or men(y). The inverse operation of subtraction gives '$x = z - y$', 'women are all people except men'. This is possible only if we assume that there is nothing in common between 'x' and 'y'. Thus, '$x + y$' is interpreted as, either 'x' or 'y'; not both. Let us consider the example: 'all members of Graduate Seminar are either Masters or Bachelors', which is expressed by the equation '$z = x+y$', where 'x' represents 'Masters', 'y' 'Bachelors', and 'z' 'members of Graduate Seminar'. Then, it is not logically possible to say: '$x = z - y$' that is, 'Masters are all members of Graduate Seminar except Bachelors'. For, there is something in common between 'x' and 'y', since 'Masters' are also 'Bachelors'. Hence Boole adopts an exclusive interpretation of 'or' expressed by '$+$' precisely in order to obtain a duality between addition and subtraction. However, the result of the exclusive interpretation of the symbol '$+$' is to bar from Boole's logical calculus the equation '$x + x = x$', a special law of an algebra of logic, which Jevons introduces.

As in arithmetical algebra, successive operation is performed by multiplication to denote the intersection of the classes 'x' and 'y'. For example, if we suppose 'x' represents 'all animals with horns'; 'y' represents 'sheep', then the logical product 'xy' represent 'sheep with horns'.

The sign of identity '$=$' indicates that the two classes between which it stands are the same and thus have the same members.

3.1.2 The Laws of Boole's Algebra of Logic

The symbols of Boole's logic are, in their use, dependent upon definite laws, 'partly agreeing with and partly differing from the laws of the corresponding symbols in the science of Algebra' (Boole 1854, p. 27). Thus, regarding the operation of logical addition, Boole states that it is commutative:

$$x + y = y + x.$$

The logical multiplication is distributive over it:

$$z(x + y) = zx + zy.$$

For example, 'European men and women' is the same as 'European men and European women'. He also implicitly implies its associativity:

$$x + (y + z) = (x + y) + z.$$

From the logical subtraction '$x - y$' (e.g. 'men excepting Asiatics), Boole infers:'$-y + x$' (e.g. excepting Asiatics, men) i.e.

$$x - y = -y + x.$$

As in arithmetical algebra, he also obtains:

$$z(x - y) = zx - zy.$$

An example of this would be:'z' represents the adjective 'white' applied to the phrase 'men except Asiatics', which is the same as to say, 'white men, except white Asiatics'.

As for logical product, since the order of the operation of selection does not affect the result, then the commutative law holds:

$$xy = yx.$$

Boole's Computational Procedure in Logic

Also, the associative law holds:
$$(xy)z = x(yz).$$

The operation of selection from classes also leads him to a law that holds in arithmetic only for 0 and 1, that is, 'the index law':
$$x^2 = x.$$

The neutral element law:
$$0\,x = 0;$$
$$1\,x = x,$$

leads him to hold that in logic '0' represents 'Nothing' and '1' represents 'Universe' or what de Morgan calls the 'universe of discourse'. The universal class '1' enables him to write: '$1-x$' for the 'contrary' of the class denoted by 'x'. For example, if 'x' refers to the class of living beings, then the expression '$1-x$' represents the universe except the 'xs or all things which are not 'xs, that is, the class of inanimate beings, the complement of 'x'. Thus, he can express the principle of non-contradiction as follows:
$$x(1-x) = 0.$$

This principle, which Aristotle described as the fundamental axiom of all philosophy, can be drawn also from the fundamental law of thought:[1]

$$x^2 = x$$
$$\therefore x - x^2 = 0$$
$$\therefore x1 - x^2 = 0$$
$$\therefore x(1-x) = 0.$$

The empty class '0' implicitly commits him to holding the law
$$x + 0 = x.$$

---

[1] Boole assigns the principle of non-contradiction to the place of just another axiom, and regards it as algebraically equivalent to the principle of idempotence (viz. the result of an act of selection, which is repeated several times, is simply equivalent to the act of the first operation). Thus, 'what has been commonly regarded as the fundamental axiom of metaphysics is but the consequence of a law of thought, mathematical in its form' (Boole 1854, p. 50).

It allows him, as well, to give an equational form to logical deductions, and to make possible its algebraic treatment. '1' and '0' may be called neutral element which are the two limits of class extension.

Boole observes that if there is the equation: '$x = y$', then whatever class 'z' may represent, there is also '$zx = zy$', which is formally the same as the algebraic law: 'If both members of an equation are multiplied by the same quantity, the products are equal'[2]. Though he does not explicitly state them, the following transposition laws hold as well:

$$\text{If } x = y, \text{ then } x + z = y + z;$$
$$\text{If } x = y, \text{ then } x - z = y - z;$$
$$\text{If } x = y, \text{ then } zx = zy.$$

Let us sum up now the laws of Boole's algebra that are explicitly or implicitly assumed:

$$x + y = y + x$$
$$x - y = -y + x$$
$$xy = yx$$
$$x + (y + z) = (x + y) + z$$
$$(xy)z = x(yz)$$
$$z(x + y) = zx + zy$$
$$z(x - y) = zx - zy$$
$$x^2 = x$$
$$x(1 - x) = 0$$
$$x + 0 = x$$
$$1\,x = x$$
$$\text{If } x = y, \text{ then } zx = zy$$
$$\text{If } x = y, \text{ then } x + z = y + z$$
$$\text{If } x = y, \text{ then } x - z = y - z.$$

---

[2] Boole goes on to note that there is a case in which the analogy with algebra does not hold: 'Suppose it true that those members of a class x which possess a certain property z are identical with those members of a class y which possess the same property z, it does not follow that the members of the class x universally are identical with the members of the class y. Hence it cannot be inferred from the equation $zx = zy$, that the equation $x = y$ is also true'. (Boole 1854, p. 36) So, the law of algebra, that sides of an equation may be divided by the same quantity has no formal equivalent here.

As a result, it is possible now that logic may be considered as a particular kind of algebra in which the quantitative symbols have only the values 0 and 1. Hence Boole concludes:

> Let us conceive, then, of an Algebra in which the symbols x, y, z, etc., admit indifferently of the values 0 and 1, and these values alone. The laws, the axioms, and the processes, of such Algebra will be identical in their whole extent with the laws, the axioms, and the processes of an Algebra of Logic. Difference of interpretation will alone divide them. Upon this principle the method of the following work is established. (Boole 1854, pp. 37–38)

As an illustration of the general principle of this algebra of logic, Boole refers to the definition of wealth due to the economist N. W. Senior: 'wealth consists of things transferable, limited in supply, and either productive of pleasure or preventive of pain'. With w standing for 'wealth', t for 'things transferable', s for 'limited in supply', p for 'productive of pleasure', r for 'preventive of pain', Boole, after omitting the conjunction 'and' regarded as superfluous, obtains the equations:

$$w = st(pr + p(1-r) + r(1-p))$$

and,

$$w = st(p(1-r) + r(1-p))$$

according to whether the 'or' in 'either productive of pleasure or preventive of pain' is interpreted inclusively or exclusively (Boole 1854, pp. 59–60).

After having set up the system of signs of the algebra and formulated the laws, Boole proceeds to state the rules of the operation of this kind of algebra that holds solely for the numbers 1 and 0, and then to build up his logical calculus.

3.2 Boole's Procedures of Computation

Boole's general method may be described as follows: he formulates a logical problem by means of an equation in which symbols have a logical sense; he operates upon the symbols without taking into account their meaning, giving them only an algebraic sense; then eventually he restores their logical sense by the procedure of interpretation. The development and elimination procedures enter into play in the intermediary step between the formulation of the logical problem and the interpretation of its solution. These formal processes of computation are viewed as 'sufficient for all the practical ends of logic' (Boole

1854, p. 130). They indeed lay the groundwork for Boole's general method in logic.

3.2.1 Development of Logical Function

Without giving any special definition of a logical function, Boole says only that in expressions, such as
$$f(x), \quad f(x,y), \quad \ldots$$
the symbol x, y, ... may be regarded as logical symbols, admitting only the values 0 and 1. Then he carries out the mechanical procedure of development that allows him to compute logical problems.

Boole defines the procedure of development as follows:

> Any function $f(x)$, in which x is a logical symbol, or a symbol of quantity susceptible only of the values 0 and 1 is said to be developed, when it is reduced to the form $ax + b(1-x)$, a and b being so determined as to make the result equivalent to the function from which it was derived. (Boole 1854, p. 72)

Thus, the definition concerns any function, logical or not; it suffices only that the variable x is restricted to the values 1 and 0. As a result, the general formula for development of a logical function can be formulated: Let us suppose

$$f(x) = ax + b(1-x).$$

In order to determine the values of a and b, we need only to substitute for x the values 1 and 0. Then, we obtain:

$$f(1) = a\,1 + b(1-1) = a$$
$$f(0) = a\,0 + b(1-0) = b.$$

If we substitute in the function, $f(1$ for a and $f(0)$ for b, then we can infer that
$$f(x) = f(1)\,x + f(0)(1-x)$$

This formula is the development of the function $f(x)$ with respect to x.

Boole goes on to develop a function involving any number of logical symbols. He first begins with a function that involves two symbols, x and y: $f(x,y)$. Considering this first as a function of x alone, and developing it by the general theorem:

$$f(x) = f(1)\,x + f(0)(1-x),$$

Boole's Computational Procedure in Logic

he can write:
$$f(x,y) = f(1,y)\,x + f(0,y)(1-x).$$
Then, regarding the result as a function of y Boole can write:
$$\begin{aligned}f(x,y) = {} & f(1,1)\,x\,y + f(1,0)\,x\,(1-y) + \\ & f(0,1)(1-x)\,y + f(0,0)(1-x)(1-y),\end{aligned}$$
which is the complete expansion of $f(x,y)$. Thus, functions involving any number of logical symbols may be developed in a similar way.

Boole states the general rule of development as follows:

> 1st. To expand any function of the symbols x, y, z.—Form a series of constituents in the following manner: Let the first constituent be the product of the symbols; change in this product any symbol z into $1-z$ for the second constituent. Then in both these change any other symbol y into $1-y$ for two more constituents. Then in the four constituents thus obtained change any other symbol x into $1-x$ for four new constituents, and so on until the number of possible changes is exhausted.
>
> 2ndly. To find the coefficient of any constituent.—If that constituent involves x as factor, change in the original function x into 1; but if it involves $1-x$ as a factor, change in the original function x into 0. Apply the same rule with reference to the symbol y, z etc.: the final calculated value of the function thus transformed will be the coefficient sought. (Boole 1854, pp. 75–76)

Accordingly, the general rule of development consists of two parts: the first introduces the *constituents* of the expansion; the second determines their respective *coefficients*. Thus, a function with three arguments has $2^3 = 8$ terms. These terms are formed by a coefficient equal to 0 or 1 and a product of the symbols, that is, the constituent. The sum of the *constituents*, multiplied each by its respective *coefficient*, is the development required. For example, we have respectively the following *constituents* and *coefficients* for a function involving three logical symbols:

$$xyz$$
$$xy(1-z)$$
$$x(1-y)z$$
$$x(1-y)(1-z)$$

$$(1-x)yz$$
$$(1-x)y(1-z)$$
$$(1-x)(1-y)z$$
$$(1-x)(1-y)(1-z)$$
$$1,1,1$$
$$1,1,0$$
$$1,0,1$$
$$1,0,0$$
$$0,1,1$$
$$0,1,0$$
$$0,0,1$$
$$0,0,0.$$

Hence the required development of the function:

$$\begin{aligned}f(x,y,z) = \ & f(1,1,1)\ xyz \\ & + f(1,1,0)\ xy(1-z) \\ & + f(1,0,1\ )x(1-y)z \\ & + f(1,0,0)\ x(1-y)(1-z) \\ & + f(0,1,1)\ (1-x)yz \\ & + f(0,1,0)\ (1-x)y(1-z) \\ & + f(0,0,1)\ (1-x)(1-y)z \\ & + f(0,0,0)\ (1-x)(1-y)(1-z).\end{aligned}$$

Boole's procedure of development corresponds to what we now call 'disjunctive normal form'. A formula of the propositional calculus is said to be in disjunctive normal form if it contains disjunction and negation as propositional connectives, applies only to propositional variables, and does not conjoin any disjunctions. Now, it turns out that in the section 'Properties of Elective

Functions' of *The Mathematical Analysis of Logic*, Boole notes that any formula $\Phi(x)$ of propositional logic containing the propositional variable x is equal to the formula
$$\Phi(1)\,x + \Phi(0)(1-x).$$
Here 1 and 0 are constants of propositional logic for truth and falsity. The quantities $\Phi(0)$ and $\Phi(1)$ are called the *moduli* of the function $\Phi(x)$. Replacing Boole's exclusive disjunction with the inclusive ∪, the formula can be written in modern notation as,
$$(\varphi(1) \cap x) \cup (\varphi(0) \cap \neg x),$$
in which the disjunctions are incompatible. It follows that Boole was aware that every formula of propositional logic is equal to a disjunctive normal form. As Boolos notices it,

> What Boole realized was that iterating this operation shows that an arbitrary propositional formula
> $$\varphi(x_1, \ldots, x_m)$$
> is equivalent to the disjunction of the 2^m formulas
> $$(\varphi(i_1, \ldots, i_m) \cap \pm x_1 \cap \ldots \cap \pm x_m),$$
> where each i_j is either 1 or 0 and $\pm x_j$ is x_j or $\neg x_j$ according as x_j is 1 or 0. Boole termed the equivalence the law of development, and called (his analogues of) the constant formulae $\varphi(i_1, \ldots, i_m)$ the moduli of $\varphi(x_1, \ldots, x_m)$. Since each modulus is equal to 1 or 0 (and it can be easily calculated which), every propositional formula
> $$\varphi(x_1, \ldots, x_m)$$
> is, as Boole saw, equivalent to the disjunction of those formulae
> $$x_1 \cap \cdots \cap \pm x_m$$
> for which the moduli $\varphi(i_1, \ldots, i_m)$ do not vanish (are not = 0). (Boolos 1998, p. 244).

Thus, in the historical notes of the propositional calculus, Church recognises that 'the full disjunctive normal form may be traced back to Boole's law of development' (Church 1956, pp. 165–166).

The disjunctive normal form has been used for the purpose of formalizing deductive logic. For instance, in *Principles of Mathematical Logic*, Hilbert and Ackermann have shown that the decision problem[3] for the propositional calculus can be worked out by using Boole's development of the disjunctive normal form of the propositional variables. According to them, 'the disjunctive normal form has the advantage of special clarity. The individual components of the disjunctive indicate the various possible cases in which the given sentential combination holds true. Thus, for example, the disjunctive normal form which belongs to $X \equiv Y$ reads $(X \mathbin{\&} Y) \cup (\overline{X} \mathbin{\&} \overline{Y})$, and this enables us to recognize that X and Y must either both be true or both be false in order for $X \equiv Y$ to be true' (Hilbert and Ackermann 1928, p. 18).

Boole comes up with the procedure of development whilst seeking a 'general method'. Although, he does not state precisely what he means by 'general method', he stresses the difference between the algebraic notation, which constitutes his basic logic, and the general method of development. This difference lies in the fact that the basic logic is restricted to operating with meaningful logical operations, whereas the general method of development applies pure formal procedures in logic. Boole wants to employ non-logical algebraic procedures in logic, so as to reinforce logical deduction. Thus, the method of development is a procedure by which, once the symbols and rules are defined, it is possible to compute, and to obtain results, without interpreting every step of the processing. It exemplifies Boole's computational procedure in logic.

There are some general principles relating to the use of the method, which describe the conditions for a valid reasoning by the means of symbols. Boole states them as follows:

> 1st. That a fixed interpretation be assigned to the symbols employed in the expression of the data; and that the laws of the combination of those symbols be correctly determined from that interpretation.
>
> 2nd. That the formal processes of solution or demonstration be conducted throughout in obedience to all the laws determined above, without regard to the question of the interpretability of the particular results obtained.
>
> 3rd. That the final result be interpretable in form, and that it be actually interpreted in accordance with that system of interpreta-

---

[3] Wilder defines a decision procedure as follows: 'given a collection C, of formulas in a theory T, a *decision problem* for C in T is the problem of finding a method-an effective procedure-by which, given any formula, we can decide in a finite number of steps whether it is in C. If such a method exists, it may be called a *decision method* or *decision procedure* for C' (Wilder 1965, p. 275).

tion which has been employed in the expression of the data. (Boole 1854, p. 68)

3.2.2 The Logical Interpretation

Boole is aware that what is of interest for a logician employing the formal procedure of the method of development is the finding of logical interpretation for the result of the processing. In this sense, he writes:

> The constituents of the expansion of any function of the logical symbols x, y, &c., are interpretable, and represent the several exclusive divisions of the universe of discourse, formed by the predication and denial in every possible way of the qualities denoted by the symbols x, y, &c. (Boole 1854, p. 81)

Thus, if the function $f(x, y)$ is considered with *constituents*,

$$xy$$
$$x(1-y)$$
$$(1-x)y$$
$$(1-x)(1-y),$$

then 'xy' will represent the class that has both the qualities expressed by x and y; '$x(1-y)$' the class having the quality x, but not the quality y; '$(1-x)y$' the class with the quality y, but not the quality x; and '$(1-x)(1-y)$' the class that does not have any of the qualities considered. These four classes described by affirmation and negation of the qualities x and y are distinct from each other, but form together the universe of discourse.

In order to determine the interpretation of any logical equation of the form $V = 0$, Boole states this rule:

> Develop the function V, and equate to 0 every *constituent* whose *coefficient* does not vanish. The interpretation of these results collectively will constitute the interpretation of the given equation. (Boole 1854, p. 83)

Regarding the *coefficients*, he gives the following canons of interpretation:

> 1st. The symbol 1, as the coefficient of a term in a development, indicates that the whole of the class which that constituent represents, is to be taken.
>
> 2nd. The coefficient 0 indicates that none of the class are to be taken.

3rd. The symbol $\frac{0}{0}$ indicates that a perfectly *indefinite* portion of the class, i.e. *some*, *none*, or *all* of its members are to be taken.

4th. Any other symbol as a coefficient indicates that the constituent to which it is prefixed must be equated to 0. (Boole 1854, p. 92)

But, in addition Boole states that

> If the solution of a problem, obtained by development, be of the form
> $$w = A + 0\,B + \frac{0}{0}C + \frac{1}{0}D,$$
> that solution may be resolved into the two following equations, viz.,
> $$w = A + v\,C$$
> $$D = 0,$$
> v being an indefinite class symbol. (Boole 1854, p. 92)

The interpretation of the first equation shows what elements are or may be a part of the composition of w, the class of things whose definition is sought; and the interpretation of the second equation shows what relations exist among the elements of the original problem, independently of w. Boole justifies the second equation by premising as a theorem that

> If a function V, intended to represent any class or collection of objects, w, be expanded, and if the numerical coefficient, a, of any constituent in its development, do not satisfy the law.
> $$a(1-a) = 0,$$
> then the constituent in question must be made equal to 0. (Boole 1854, p. 90)

After having given the proof of the theorem (Boole 1854, pp. 90–91), Boole then shows generally that

> Any constituent whose coefficient is not subject to the same fundamental law as the symbols themselves must be separately equated to 0. (Boole 1854, p. 91)

According to him, the usual form under which such coefficients occur is '$\frac{1}{0}$', which does not satisfy the fundamental law above referred to.

As an example for the interpretation of logical equations, let us take the definition of 'clean beasts' as laid down in the Jewish law, viz., 'clean beasts

Boole's Computational Procedure in Logic

are those which both divide the hoof and chew the cud'. The proposition is represented by the equation

$$x = yz,$$

where x represents the clean beasts, y beasts dividing the hoof and z beasts chewing the cud. In order to determine the relation, in which 'beasts chewing the cud' stands to 'clean beasts' and 'beasts dividing the hoof', we divide by y:

$$z = \frac{x}{y}$$

But being a division this equation is not in a logically interpretable form, and thus its right hand side,

$$\frac{x}{y}$$

requires to be developed according to the formula for expansion of logical functions into constituents that is,

$$\frac{x}{y} = f(x,y)$$
$$= f(1,1)\, xy + f(1,0)\, x(1-y)+$$
$$f(0,1)\, (1-x)y + f(0,0)\, (1-x)(1-y)$$

Thus, we have the following coefficients:

$$f(1,1) = 1$$
$$f(1,0) = \frac{1}{0}$$
$$f(0,1) = \frac{0}{1}$$
$$f(0,0) = \frac{0}{0}.$$

Hence,

$$z = xy + \frac{1}{0}\, x(1-y) + \frac{0}{1}\, (1-x)y + \frac{0}{0}\, (1-x)(1-y).$$

The expression may be written as,

$$z = xy + 0(1-x)y + \frac{0}{0}\, (1-x)(1-y) + \frac{1}{0}\, x(1-y),$$

which is of the form

$$w = A + 0B + \frac{0}{0}\, C + \frac{1}{0}\, D.$$

Hence, according to Boole's canons of interpretation stated above, the developed expression may be rewritten into the two following equations:

$$z = xy + v(1-x)(1-y)$$
$$x(1-y) = 0$$

Boole interprets the first equation as saying:

> Beasts which chew the cud consist of all clean beasts (which also divide the hoof) $[xy]$, together with an indefinite remainder (some, none, or all) [indicated by v] of unclean beasts which do not divide the hoof $[v(1-x)(1-y)]$. (Boole 1854, p. 87)

The second equation may be interpreted as: 'There are no clean beasts which do not divide the hoof' $[x(1-y) = 0]$. This is an independent relation which conveys a meaningful piece of information. It is a bonus that the procedure of development brings out.

This shows the fecundity and the ingenuity of Boole's procedure of computation, which yields a logically interpretable solution. The procedure seems to survive in modern abstract algebra, which embodies an infinity of systems, only some of which are logically interpretable.

3.2.3 Elimination of Logical Symbols

As the middle term is eliminated in a syllogism in order to unfold the conclusion which was implicit in the premises, Boole uses the procedure of elimination to get rid of logical symbols which are not expected to appear in the conclusion. He does so by means of the algebraic notation, and this in a more general and systematic way than the traditional argument. Thus, Boole claims that the object of the method is to eliminate any number of symbols from any number of logical equations, and to present in the result the relations which remain. By contrast with arithmetical algebra, in which we are able to eliminate one symbol from two equations, two symbols from three equations, and generally $n-1$ symbols from n equations, Boole points out that in a logical system from a single equation an indefinite number of such symbols may be eliminated independently of the number of equations. He states the theorem of the method of elimination as follows:

> If $f(x) = 0$ be any logical equation involving the class symbol x, with or without other class symbols, then will the equation $f(1)f(0) = 0$ be true, independently of the interpretation of x from the above equation. In other words the elimination of x from any

Boole's Computational Procedure in Logic 61

given equation, $f(x) = 0$, will be effected by successively changing in that equation x into 1, and x into 0, and multiplying the two resulting equations together. Similarly the complete result of the elimination of any class symbols, x, y &c., from any equation of the form $V = 0$, will be obtained by completely expanding the first member of that equation in constituents of the given symbols, and multiplying together all the coefficients of those constituents, and equating the product to 0. (Boole 1854, p. 101)

Boole proves this theorem in three different ways. Let us consider only the first proof, which is entirely algebraic. He develops the equation $f(x) = 0$ to have

$$f(1)\, x + f(0)\, (1-x) = 0,$$

which can be rewritten in the form

$$(f(1) - f(0))\, x + f(0) = 0.$$

From this we can infer that

$$x = \frac{f(0)}{f(0) - f(1)}$$

which gives

$$\begin{aligned}1 - x &= 1 - \frac{f(0)}{f(0) - f(1)} \\ &= \frac{f(0) - f(1)}{f(0) - f(1)} - \frac{f(0)}{f(0) - f(1)} \\ &= -\frac{f(1)}{f(0) - f(1)}.\end{aligned}$$

Now, according to the fundamental law of his algebra

$$x(1-x) = 0$$

we can substitute these expressions for x and $1-x$

$$\left(\frac{f(0)}{f(0) - f(1)}\right) - \left(\frac{f(1)}{f(0) - f(1)}\right) = 0$$

and then obtain as a result

$$-\frac{f(0)f(1)}{(f(0) - f(1))^2} = 0;$$

since where $\frac{a}{b} = 0$, $a = b \times 0 = 0$, hence

$$f(0)f(1) = 0,$$

the equation sought.

The proof leads to the rule of elimination of any symbol from a proposed equation:

> The terms of the equation having been brought, by transposition if necessary, to the first side, give to the symbol successively the values 1 and 0, and multiply the resulting equations together. (Boole 1854, p. 103)

As an example of the procedure of elimination, let us resume N. W. Senior's definition of wealth: 'wealth consists of things transferable, limited in supply, and either productive of pleasure or preventive of pain'. With w standing for wealth, t things transferable, s limited in supply, p productive of pleasure, r preventive of pain, Boole, after assuming the inclusive interpretation of the 'or' in 'either productive of pleasure or preventive of pain', obtains this equation:

$$w = st(pr + p(1-r) + r(1-p))$$

or

$$w = st(p + r(1-p)),$$

Then, Boole notes that from this equation we can eliminate any unwanted symbols and express the result by the procedure of development and interpretation. Let us see what the expression for w, wealth, would be if the element r, standing for preventive of pain, were eliminated. By bringing the terms of the equation to the first side, we have

$$w - st(p + r - rp) = 0.$$

Considering $r = 1$, the left hand side of the equation yields '$w - st$', and considering $r = 0$ it yields '$w - stp$'. Then applying the rule of elimination, that is, $f(1)f(0) = 0$ we get

$$(w - st)(w - stp) = 0,$$

or

$$w - wstp - wst + stp = 0,$$

which can be rewritten as,

$$w(1 - stp - st) + stp = 0,$$

Boole's Computational Procedure in Logic

or
$$w(1 - stp - st) = -stp.$$

It is then easy to infer that
$$w = \frac{stp}{st + stp - 1};$$

and according to the formula for expansion of logical functions into constituents, that is,

$$\begin{aligned}
f(s,t,p) = {} & f(1,1,1)\ stp \\
& + f(1,1,0)\ st(1-p) \\
& + f(1,0,1)\ s(1-t)p \\
& + f(1,0,0)\ s(1-t)(1-p) \\
& + f(0,1,1)\ (1-s)tp \\
& + f(0,1,0)\ (1-s)t(1-p) \\
& + f(0,0,1)\ (1-s)(1-t)p \\
& + f(0,0,0)\ (1-s)(1-t)(1-p)
\end{aligned}$$

we develop the right hand side of the equation, that is,

$$\frac{stp}{st + stp - 1},$$

because it is a division which is not logically interpretable. Thus, we have the following coefficients:

$$f(1,1,1) = 1,$$
$$f(1,1,0) = \frac{0}{0}$$
$$f(1,0,1) = -\frac{0}{1}$$
$$f(1,0,0) = -\frac{0}{1}$$
$$f(0,1,1) = -\frac{0}{1}$$
$$f(0,1,0) = -\frac{0}{1}$$
$$f(0,0,1) = -\frac{0}{1}$$
$$f(0,0,0) = -\frac{0}{1}.$$

Hence,

$$w = 1\,stp + \frac{0}{0} st(1-p) - \frac{0}{1} s(1-t)p - \frac{0}{1} s(1-t)(1-p) - \frac{0}{1}(1-s)tp -$$
$$\frac{0}{1}(1-s)t(1-p) - \frac{0}{1}(1-s)(1-t)p - \frac{0}{1}(1-s)(1-t)(1-p).$$

According to the 4th Rule of interpretation the terms whose coefficients are $-\frac{0}{1}$ must vanish, therefore we obtain

$$w = stp + \frac{0}{0} st(1-p),$$

which is interpreted as,

> Wealth consists of all things limited in supply, transferable, and productive of pleasure, and an indefinite remainder of things limited in supply, transferable, and not productive of pleasure. (Boole 1854, p. 107)

From all this, it can be said that the Boole's mechanical procedures of computation operate upon the basis that logical derivations can be carried out by means of symbols satisfying the laws of an algebra for *1* and 0 . He handles logical derivations almost exclusively according to algebraic laws, regardless the direct logical interpretation of the symbols. That is why Boole is often criticised for introducing uninterpretable symbols such as subtraction and division into his computational procedure (see subsection **3.6.1**). However, it should be stressed that it is quite important, in the nineteenth century, that Boole was already aware that symbols may be employed, without being 'locked in' to a determinate meaning. Nowadays, such an idea corresponds to the method of 'disinterpretation'; i.e. a procedure which is formulated purely in terms of syntactical manipulation.

3.3 Boole's 'Primary Propositions'

In what follows, I shall first be concerned with the general method for the expression of 'primary propositions'. Then I shall show how the principles and the processing of his computational procedure can be applied to them. In doing so, I shall describe Boole's procedure of computation called the method of reduction, which allows one to reduce any systems of propositions to an equivalent single one, to which the procedures of computation described above may be immediately applied.

3.3.1 The Expression of Primary Propositions

A primary proposition consists of two terms, namely the subject and the predicate between which a relation is asserted. Boole shows how the relations between these terms are to be expressed symbolically. Thus, if both subject and predicate of a proposition are universal, e.g. as where 'All fixed stars are suns' is interpreted, so as to imply 'All suns are fixed stars', then its expression will be

$$x = y.$$

Here, x represents 'fixed stars' and y 'suns' and the formula expresses a relation of identity between two classes, 'Fixed stars' and 'Suns'. It follows as a rule that 'when both subject and predicate of a proposition are universal, form the separate expressions for them, and connect them by the sign $=$'. (Boole 1854, p. 59)

Let us take the case in which the predicate is particular, that is, an affirmative universal proposition, e.g. 'All men are mortal.' The proposition means that 'All men are some mortal beings' and the expression of the predicate 'some mortal beings' is required. Since in his notation there is no place for individual variables or quantifiers to bind them (i.e. universal and existential operators), Boole expresses the proposition by introducing a special symbol 'v' to express 'some' as it occurs in 'some mortal beings'. So 'vx' represents

> A class indefinite in every respect but this, viz., that some of its members are mortal beings, and let x stand for 'mortal beings,' then will vx represent 'some mortal beings.' Hence if y represents men, the equation sought will be $y = vx$. (Boole 1854, p. 61)

It follows that the expression of the proposition 'All men are some mortal beings' is

$$y = vx.$$

Thus, Boole derives a rule for expressing an affirmative universal proposition whose predicate is particular: 'express as before the subject and the predicate, attach to the latter the indefinite symbol v, and equate the expressions' (Boole 1854, p. 61). Later, I will discuss the inconsistent treatment which Boole gives to the symbol 'v' and the problems and misunderstandings it causes (see subsection **3.6.2**).

Considering the case of universal negative propositions, e.g. 'No men are perfect beings', Boole notes that

> We do not speak of a class termed 'no men', and assert of this class that all its members are 'perfect beings'. But we virtually make an assertion about 'all men' to the effect that they are '*not perfect beings*'. (Boole 1854, p. 62)

Thus, the true meaning of the proposition is : 'All men are not perfect.' With y representing 'men', and x 'perfect beings,' we have

$$y = v(1 - x).$$

Hence the rule: 'to express any proposition of the form "No xs are ys," convert it into the form "All xs are not ys.", and then proceed as in the previous case'. (Boole 1854, p. 63)

Finally come the cases in which the subject of the proposition is particular, e.g. 'Some men are wise' and 'Some men are not wise'. In the latter case, Boole remarks that the negative 'not' may be referred to the predicate 'wise',

> ... for we do not mean to say that it is not true that 'Some men are wise,' but we intend to predicate of 'some men' a want of wisdom.

Thus, the form of the proposition is 'Some men are not-wise.' With y representing 'men', x 'wise', and introducing v as the symbol of a class indefinite

Boole's Computational Procedure in Logic 67

in all respect but this, that it contains some individuals of the class to whose expression it is prefixed' (Boole 1854, p. 63), we have

$$vy = v(1-x).$$

The particular affirmative 'Some men are wise' is expressed as,

$$vx = vy.$$

As a result, Boole embodies all these cases in the general rule for the symbolical expression of primary propositions:

> 1st. If the proposition is affirmative, form the expression of the subject and that of predicate. Should either of them be particular, attach to it the indefinite symbol v, and then equate the resulting expressions.
>
> 2ndly. If the proposition is negative, express first its true meaning by attaching the negative particle to the predicate, then proceed as above. (Boole 1854, p. 63)

It is then easy to infer the symbolical expression of the four fundamental types of reasoning as follows: (Boole 1854, p. 228)

A	All Ys are Xs	$y = vx$
E	No Ys are Xs	$y = v(1-x)$
I	Some Ys are Xs	$vy = vx$
O	Some Ys are not-Xs	$vy = tv(1-x)$

There is what Boole calls 'the modern extension of Aristotelian logic' due to the introduction by de Morgan of contrary terms, which has led to the enlargement of the four traditional forms to eight. Accordingly, Boole expresses the four remaining forms by substituting $1-y$ for y in the above **A**, **E**, **I**, **O** forms.

3.3.2 Reduction of Systems of Propositions

Since the procedure of elimination is applicable only to logical equations taken individually, Boole introduces another formal device, which may be applied to systems of propositions. The method is a technique of reducing a system of propositions to an equivalent single proposition, which may then be interpreted as it has been described above. Boole carries out a first method called, the 'method of indeterminate multipliers'[4], by which each equation after the first

---
[4] This method is presented in the last pages of *The Mathematical Analysis of Logic*, pp. 78–81.

is multiplied by an arbitrary constant and the equations then added. But these indeterminate multipliers have the inconvenience of complicating the procedure of elimination and development. Hence, Boole adopts a second method, which does not require the introduction of arbitrary constants, and thereby is preferable. The rule of this method is stated as follows:

> 1st. That any equations which are of the form $V = 0$, V satisfying the fundamental law of duality $V(1-V) = 0$, may be combined together by simple addition.
>
> 2nd. That any other equation of the form $V = 0$ may be reduced, by the process of squaring, to a form in which the same principle of combination by mere addition is applicable. (Boole 1854, p. 123)

Subsequently, Boole determines what equations in the expression of propositions conform to the first, and what to the second group. This leads him to consider what he calls 'the great leading types of propositions symbolically expressed' (Boole 1854, p. 64). So an example of an equation of the first group is

$$X = vY.$$

It satisfies the law of duality and is the symbolical expression of propositions of which the subject is universal and the predicate particular. Boole writes it in the equivalent form

$$X(1-Y) = 0$$

by eliminating the symbol 'v'. This is done by writing the equation '$X = vY$' in 'zero form', that is, '$X - vY = 0$'. Thus, in making $v = 0$, we have $X = 0$; and $v = 1$, $X - Y = 0$. Then, according to the rule of elimination, that is, $f(1)f(0) = 0$, we equate their product to 0 and obtain '$X(X - Y) = 0$', or '$X(1-Y) = 0$'.

An example of an equation of the second group is

$$X = Y.$$

It expresses propositions of which both terms are universal. It is written in 'zero form', that is,

$$X - Y = 0.$$

Then, Boole squares both sides of the equation to obtain

$$X - 2XY + Y = 0,$$

or

$$X(1-Y) + Y(1-X) = 0,$$

Boole's Computational Procedure in Logic

which satisfies the law of duality and can be logically interpreted.

Considering again the equation

$$X = Y,$$

Boole remarks that we may arrive at the same equation in a different manner. Thus, the equality '$X = Y$' is equivalent to the two equations

$$X = vY$$
$$Y = vX,$$

'for to affirm that Xs are identical with Ys is to affirm both that All Xs are Ys and that All Ys are Xs.' Now eliminating v, these equations give

$$X(1 - Y) = 0$$
$$Y(1 - X) = 0,$$

of which the sum is equivalent to the preceding equation that is,

$$X(1 - Y) + Y(1 - X) = 0.$$

Finally, Boole considers a proposition of which both terms are particular, which is an example of an equation of the second group. The form of the equation is

$$vX = vY.$$

By transposing the second member to the first side, and squaring the resulting equation according to the rule, we have the equation

$$vX(1 - Y) + vY(1 - X) = 0.$$

Boole embodies the above considerations in a rule:

> The equations being so expressed as that the terms X and Y in the following typical forms obey the law of duality, change the equations
>
> $$\begin{array}{lll} X = vY & \text{into} & X(1-Y) = 0 \\ X = Y & \text{into} & X(1-Y) + Y(1-X) = 0 \\ vX = vY & \text{into} & vX(1-Y) + vY(1-X) = 0. \end{array}$$
>
> Any equation which is given in the form $X = 0$ will not need transformation, and any equation which presents itself in the form $X = 1$ may be replaced by $1 - X = 0$, as appears from the second of the above transformations. (Boole 1854, p. 124)

3.4 Systematisation and Generalisation of Aristotelian Logic

Boole's computational procedure provides a general and systematic method which enables him to perform the operations of Conversion and Syllogism, and thereby to cast Aristotelian logic in the form of a computation. He describes his aim as follows:

> The course which I design to pursue is to show how these processes of Syllogism and Conversion may be conducted in the most general manner upon the principles of the present treatise, and, viewing them thus in relation to a system of Logic, the foundations of which, it is conceived, have been laid in the ultimate laws of thought, to seek to determine their true place and essential character. (Boole 1854, p. 228)

In traditional theory, Syllogism appears as various processes without a veritable corpus, for the procedures are handled case by case without a systematic unity. The different types of Conversion use different forms of processing strictly distinguishable from those of Syllogism. The disparate technique of the art of the mnemonic verses particularly convinces Boole that the scholastic logic

> Is not a science, but a collection of scientific truths, too incomplete to form a system of themselves, and not sufficiently fundamental to serve as the foundation upon which a perfect system may rest. (Boole 1854, p. 241)

Boole's Computational Procedure in Logic

Thus, in order to lay down the foundation of a better system he aims to show how readily the rules of the logical operation of Conversion and Syllogism can be followed when performed with algebraic symbolism.

However, it should be mentioned that Boole's criticism of the traditional theory does not apply to the Aristotelian system itself. Moreover, modern axiomatisations of Aristotelian logic make a distinction between Aristotelian system and the traditional theory. For instance, Lukasiewicz sets up Aristotelian theory of the syllogism as an axiomatic deductive system in which the reduction of the other syllogistic moods to those of the first figure is a part of the system. As well, in Lear's eyes, Aristotle had a unified and coherent notion of logical consequence; and he even attempts to demonstrate the completeness and compactness of Aristotle's logical programme. (Lear 1980, pp. 15–33)

Nonetheless Boole provides a valuable alternative approach to Aristotelian logic, which constitutes an advance in the history of logic.

3.4.1 Boole's Laws of Conversion

From the symbolical expressions of the four fundamental propositions given above, Boole draws his laws of Conversion. Conversion consists, sometimes, in a simple transposition of the terms of a proposition, without altering their quality, as when we change Some Ys are Xs into Some Xs are Ys, that is,

$$vy = vx \text{ into } vx = vy.$$

Let us consider the universal affirmative all Ys are Xs which is represented by the equation,

$$y = vx.$$

Here the symbol v must be eliminated, according to the rule of elimination which consists of bringing the terms of the equation to the first side, giving to the unwanted symbol successively the values 1 and 0, and multiplying the resulting equations together. Thus, from the given equation we have

$$y - vx = 0,$$

then making $v = 0$ gives $y = 0$; and making $v = 1$ gives $y - x = 0$. By multiplying the resulting equations we obtain

$$y(y - x) = 0$$

and

$$y - yx = 0,$$

so,
$$y(1-x) = 0,$$
which gives by solution with reference to $1-x$,
$$1-x = \frac{0}{y}.$$
Since the right hand side of this equation is a division and is not logically interpretable, it must be developed according to the formula for expansion of logical functions into constituents, that is
$$f(y) = f(1)\,y + f(0)\,(1-y).$$
Thus, we have the following coefficients
$$f(1) = \frac{0}{1}$$
$$f(0) = \frac{0}{0}$$
which enable one to form the constituents of the equation and to obtain its development:
$$1-x = \frac{0}{1}y + \frac{0}{0}(1-y).$$
According to the Rule of interpretation, the term whose coefficient is $\frac{0}{1}$ must vanish; and the symbol $\frac{0}{0}$ may be replaced by the symbol v. Hence we have:
$$1-x = v(1-y);$$
which is interpreted as: All not-Xs are not-Ys. This is an example of 'negative conversion'.

Boole takes it that these examples demonstrate that Conversion is a particular application of a much more general procedure in logic. In Boole's eyes, such a procedure

> ...has for its object the determination of any element in any proposition, however complex, as a logical function of the remaining elements. Instead of confining our attention to the subject and predicate, regarded as simple terms, we can take any element or any combination of elements entering into either of them; make that element or that combination, the 'subject' of a new proposition; and determine what its predicate shall be, in accordance with the data afforded to us. (Boole 1854, p. 230)

Boole's Computational Procedure in Logic 73

Furthermore, Boole remarks that even the simple forms of the propositions enumerated above afford some ground for the application of such a method, beyond what the traditional laws of conversion offer. Thus, the universal affirmative proposition, all Ys are Xs, which is represented by the equation,

$$y = vx,$$

yields the following:

$$y(1-x) = 0$$

which is the first proposition obtained by eliminating the symbol v according to the rule of elimination. Boole interprets it as follows:

There are no Ys that are not-Xs. (Boole 1854, p. 230)

From the above equation we can infer that

$$y = \frac{0}{1-x},$$

which gives by solution with reference to $1-y$,

$$\begin{aligned}1-y &= 1 - \frac{0}{1-x} \\ &= \frac{1-x}{1-x} - \frac{0}{1-x}\end{aligned}$$

hence

$$1-y = \frac{1-x}{1-x}.$$

This equation, as it stands, cannot be interpreted, so its right hand side

$$\frac{1-x}{1-x}$$

must be developed according to the formula for expansion of logical functions into constituents, that is,

$$f(x) = f(1)x + f(0)(1-x).$$

Thus, we have the following coefficients:

$$f(1) = \frac{0}{0}$$
$$f(0) = 1$$

which then enable us to form the second proposition:

$$1 - y = \frac{0}{0} x + (1 - x),$$

which Boole interprets as

> Things that are not-Ys include all things that are not-Xs, and an indefinite remainder of things that are Xs. (Boole 1854, p. 230)

In the same way, from the first equation

$$y(1 - x) = 0,$$

we can infer that

$$y - yx = 0,$$

hence,

$$x = \frac{y}{y}.$$

This equation also is not in an interpretable form and thus the right hand side requires to be developed according to the formula for expansion of logical functions into constituents, that is,

$$f(y) = f(1)\, y + f(0)\, (1 - y),$$

which gives as coefficients:

$$f(1) = 1$$
$$f(0) = \frac{0}{0}.$$

Therefore, we obtain the third proposition:

$$x = y + \frac{0}{0}(1 - y),$$

which Boole interprets as

> Things that are Xs include all things that are Ys, and an indefinite remainder of things that are not-Ys. (Boole 1854, p. 230)

What Boole shows is that we can model the procedure of Conversion within his algebra of logic so that to render it more general and systematic.

Boole's Computational Procedure in Logic 75

3.4.2 Boole's Solution of Syllogism

The mood b**A**rb**A**r**A** readily shows the nature of syllogism. Let us consider the propositions,

$$\text{All } X\text{s are } Y\text{s}$$
$$\text{All } Y\text{s are } Z\text{s}.$$

From these premises we may draw the conclusion,

$$\text{All } X\text{s are } Z\text{s}.$$

We have here what is called a *syllogistic inference*. The terms 'X' and 'Z' are the extremes, and 'Y' is the middle term, Boole defines generally the function of the syllogism as follows:

> Given two propositions... and involving one middle term or common term Y, which is connected in one of the propositions with an extreme X, in the other with an extreme Z; required the relation connecting the extremes X and Z. (Boole 1854, p. 231)

Hence, in his eyes, it is easy to use a computational procedure to give a more general conclusion of Syllogism. Thus, he worked out syllogisms by the means of the processes of elimination and reduction of systems of propositions.

Let us take the above syllogism b**A**rb**A**r**A** as an example. The premises are expressed in the symbolic language as follows,

$$x = vy,$$
$$y = vz.$$

By substitution we have

$$x = vvz.$$

Boole's treatment of 'v' is such that $vvz = vz$, hence $x = vz$ which means

$$\text{All } X\text{s are } Z\text{s}.$$

We could have used the method of reduction and elimination as well. Thus, when the symbol v is eliminated in both equations, we obtain

$$x(1 - y) = 0,$$
$$y(1 - z) = 0.$$

Then adding these equations, and eliminating y, in accordance with the rules of reduction and elimination, we have

$$x(1 - z) = 0.$$

From this equation we can infer

$$x = \frac{0}{1-z}.$$

The right hand side of this equation is uninterpretable as it stands, and so must be developed according to the formula for expansion of logical functions that is,

$$f(z) = f(1)z + f(0)(1-z)$$

which gives the following coefficients:

$$f(1) = \frac{0}{0}$$
$$f(0) = \frac{0}{1}.$$

Hence we have

$$x = \frac{0}{0}z,$$

which is interpreted into

All Xs are Zs.

Next we consider as an example c**ElArEn**t which belongs to the mood **EAE** in the first figure. The premises,

No Ys are Xs
All Zs are Ys

are expressed as follows,

$$y = v(1-x)$$
$$z = vy.$$

When the symbol v is eliminated in both equations we have

$$xy = 0$$
$$z(1-y) = 0.$$

In accordance with the procedure of reduction and elimination, we add these equations, and eliminate the unwanted symbol y to have

$$zx = 0,$$

Boole's Computational Procedure in Logic

from which we can infer

$$z = \frac{0}{x}.$$

This equation being uninterpretable, we must develop its right-hand side '$\frac{0}{x}$' according to the formula for expansion of logical functions that is,

$$f(x) = f(1)\,x + f(0)\,(1-x),$$

which gives as coefficients

$$f(1) = \frac{0}{1}$$
$$f(0) = \frac{0}{0}.$$

Therefore, we have

$$z = \frac{0}{0}\,(1-x),$$

which is interpreted into

No Zs are Xs.

The same computational procedure may be used to work out any sort of syllogism whatsoever.

It is worth emphasising the advantages of the implementation of Boole's computational procedure as compared with the Aristotelian syllogism. First, it allows us to deal with propositions involving several terms, whereas Aristotelian logic deals only with propositions with two terms. Secondly, it enables us to analyse all possible combinations, and thus to draw many legitimate conclusions. In order to illustrate this point, let us take the syllogism cElArEnt, as an example. From the premises,

No Ys are Xs
All Zs are Ys,

Aristotelian logic draws only one conclusion:

No Zs are Xs.

It is beyond question that this conclusion is entailed by the premises. However, there are others which are also satisfactory but concealed by the Aristotelian inference. Now, since the given syllogism embodies three terms, I shall not present separately the two premises; instead I shall consider the eight possible combinations of the three terms by using Boole's procedure of development of logical functions into constituents. Thus, the constituents which are inconsistent with one or the other of the premises

$$yx = 0$$
$$z(1 - y) = 0$$

are

$$xyz$$
$$xy(1 - z)$$
$$xz(1 - y)$$
$$z(1 - x)(1 - y).$$

Here the first premise '$yx = 0$' tells us that the constituents 'xyz' and '$xy(1-z)$' are empty; therefore they are false and must be eliminated. Similarly, the second premise '$z(1 - y) = 0$' eliminates the constituents '$xz(1 - y)$' and '$z(1 - x)(1 - y)$'. The remaining constituents are consistent with one or the other of the premises and therefore true:

$$x(1 - y)(1 - z) = 1$$
$$yz(1 - x) = 1$$
$$y(1 - x)(1 - z) = 1$$
$$(1 - x)(1 - y)(1 - z) = 1.$$

Regarding the constituents which are inconsistent with the premises, if we add the first and the third one, that is,

$$xyz + xz(1 - y) = 0,$$

then we obtain the traditional conclusion of the syllogism c**ElAr**Ent,

$$zx = 0$$

or

No Zs are Xs.

But there remains a certain number of other allowed combinations, which are as many legitimate conclusions of the premises. For instance, the second equation of the constituents, which are inconsistent with the premises, that is,

$$xy(1 - z) = 0$$

tells us that Xs that are Ys, but not Zs do not exist; and the first equation of the constituents which are consistent with the premises, that is,

$$x(1-y)(1-z) = 1,$$

says that there are Xs that are neither Ys nor Zs.

Here the fruitfulness of Boole's computational procedure is again exhibited. The procedure is not bound up with the traditional syllogistic conclusions in that it brings out all the possible inferences from the premises. It has the advantage of special clarity, for it indicates the various possible conclusions which are entailed by the premises.

There is an interesting connection between the decision procedure and the way in which the above procedure is handled. As Tarski puts it, a decision procedure is like 'a recipe, which tells one what to do at each step so that no intelligence is required to follow it; and the method can be applied by anyone so long as he is able to read and follow directions' [5] (Wilder 1965, p. 275).

For instance, regarding the propositional calculus, the procedure of building up a truth-table is a decision procedure for validity in that one can apply it to any formula, to determine whether it is valid, and it always gives the answer in a finite number of steps. Likewise, a syllogism has been worked out above by using Boole's formula for expanding logical functions, in which the symbols $[0, 1]$ are intended to indicate the falsity or truth of the constituents. One can test a syllogism by enumerating all possible combinations of terms, and then eliminating those that are not consistent with the premises. Thus, Boole's procedure of development virtually allows the use of 'matrix analysis',[6] a sort of truth-table method for working out logical problems, and that is exactly what one does when one considers the case of a propositional function being false or being true. Hence, although its first explicit application to truth-value propositions came later, the truth-table method was already implicit in Boole's procedure of development (see also below subsection **3.5.1**).

---

[5]Thus, it should be even possible to build up a machine that would help to perform the operation, and it has been done by Boole's followers such as, Jevons and Venn.

[6]Jevons has been often viewed as the first who makes use of matrix analysis in carrying out his method called 'combinatorial logic'. But, as Venn highlights, in 1811 Semler already insisted on the procedure of enumerating all possible combinations of terms, then eliminating those that are not consistent with the premises, which is almost identical with Jevons's method (Venn 1894, p. 415). Furthermore, as Gardner puts it, 'if by "matrix method" we mean nothing more than recognition of the alternate possible combinations of truth values for a given binary function, then this recognition goes all the way back to the ancient Stoic-Megaric school. A truth table for material implication, for example, is given by Sextus Empiricus to define the meaning of a conditional statement as it was understood by Philo of Megara' (Gardner 1982, p. 103).

Although the above procedure was not used by Boole himself to work out syllogisms, it is implied in his method of expansion. Moreover, it afforded ground for a further development. Thus Jevons devised a method which consists of building up a table that exhausts all possible combinations of given syllogistic premises, which he called the 'logical alphabet'. Then each premise forced him to eliminate certain lines of this 'truth table'. Finally he analysed the remaining lines that are all consistent with the premises to find out the relation connecting the two extremes. He called this procedure 'combinatorial logic'.

Venn also made use of the same method, which he described as follows: 'we take the given premises, break them up into fragments, and then put these fragments, or a part of them, together in some other arrangement in order to build up the structure required' (Venn 1894, p. 401). Thus, on behalf of Boole he claimed to have substituted a system of polytomy in the place of the old dichotomy (see subsection **4.1.2**).

This interesting development towards the use of the truth-table method leads us to the implementation of Boole's computational procedure in the expression of propositional calculus.

3.5 Boole's 'Secondary Propositions'

Secondary propositions are those which 'concern or relate to other propositions considered as true or false'. They express judgments concerning the truth or falsehood of other propositions: the conditional 'If the sun shines the day will be fair' is interpreted as expressing a judgment, which concerns a relation of dependence between the proposition, 'the day will be fair,' and the proposition, 'the sun shines', to the effect that the truth of the former depends on the truth of the latter. Boole also considers as secondary propositions 'all those propositions which assert truth or falsehood of propositions'. In effect therefore Boole's treatment of secondary propositions amounts to a calculus of propositions.

Boole's analysis of secondary propositions occurs in the chapter on Hypotheticals of *The Mathematical Analysis of Logic,* and in the chapters XI, XII of *The Laws of Thought*. He carries out two different approaches. In the first publication the resulting calculus of propositions is based upon the idea of 'cases' or 'circumstances'. But in *The Laws of Thought* he retreats from the idea of a 'universe of cases' to that of 'times for which a proposition is true.' I shall discuss these two developments of propositional calculus whilst emphasising the calculus of hypotheticals as it stands in *The Mathematical Analysis of Logic*. For the earlier approach to propositional logic would make him the founding father of modern logic.

3.5.1 Calculus of Hypothetical Propositions and 'Cases'

In the *Mathematical Analysis of Logic*, Boole interprets the symbol 1 in secondary propositions as the universe of 'cases' or 'conjunctures of circumstances'. Boole writes

> To the symbols X, Y, Z, representative of propositions, we may appropriate the elective symbols x, y, z, in the following sense. The hypothetical Universe, 1, shall comprehend all conceivable cases and conjunctures and circumstances. The elective symbol x attached to any subject expressive of such cases shall select those cases in which the Proposition X is true, and similarly for Y and Z. If we confine ourselves to the contemplation of a given Proposition X, and hold in abeyance any other consideration, then two cases only are conceivable, viz. first that the given Proposition is true, and secondly that it is false. As these two cases together make up the Universe of the Proposition, and as the former is determined by the elective symbol x, the latter is determined by the symbol $1 - x$. (Boole 1847, p. 49)

Thus, when 'X', 'Y', 'Z', etc. represent propositions, the hypothetical universe '1' 'comprehends all conceivable cases and conjuncture of circumstances'. The elective symbols 'x', 'y', 'z', etc. corresponding to the propositions 'X', 'Y', 'Z', etc. select, from any subject 'expressive of such cases,' those cases in which the respective propositions are true. Thus, if the propositions 'X' and 'Y' are combined, then the total number of conceivable cases are shown in the following scheme.

	Combinations of Propositions	*Combinations of Circumstances*
1st	X True, Y True	xy
2nd	X True, Y False	$x(1-y)$
3rd	X False, Y True	$(1-x)y$
4th	X False, Y False	$(1-x)(1-y)$

In a similar way if three propositions are considered, then there are eight cases in the Universe. In general, for n propositional variables, the hypothetical Universe is defined relative to these propositional variables to be the class consisting of 2^n cases.

In the above scheme if we consider the sum of the two first cases,

$$xy + x(1-y)$$

then we have

$$x,$$

the elective symbol for the general case of X being true independently of Y; and the sum of the two last cases

$$(1-x)y + (1-x)(1-y)$$

yields

$$1-x,$$

the elective symbol for the general case of X being false. The sum of the elective expressions representing any number of cases will be 1.

Boole expresses the truth of a proposition X, by saying that 'the elective symbol $1-x$ selects those cases in which the proposition X is false. But, if the proposition is true, there are no such cases in its hypothetical Universe, therefore

$$1-x=0$$

so

$$x=1.$$

Similarly the expression of a false proposition is

$$x=0$$

He says that,

> In every case, having determined the elective expression appropriate to a given Proposition, we assert the truth of that proposition by equating the elective expression to unity, and its falsity by equating the same expression to 0. (Boole 1847, p. 51)

The following rule is then stated:

> Consider what are those distinct and mutually exclusive cases of which it is implied in the statement of a given Proposition, that some one of them is true, and equate the sum of their elective expressions to unity. This will give the equation of the given Proposition. (Boole 1847, p. 52)

Boole's Computational Procedure in Logic

As an example to illustrate the rule, consider the two propositions, X and Y, that are simultaneously true. Boole employs the multiplication sign, which affirms the first case, and denies the other three. Therefore the only case to be considered is

1st	X True, Y True	xy

According to the rule, the equation sought is

$$xy = 1.$$

When we develop the first member of the equation according to Boole's procedure of development of logical functions into constituents, that is,

$$f(x,y) = f(1,1)xy + f(1,0)x(1-y) + f(0,1)(1-x)y + f(0,0)(1-x)(1-y),$$

then we have the following coefficients:

$$f(1,1) = 1$$
$$f(1,0) = 0$$
$$f(0,1) = 0$$
$$f(0,0) = 0.$$

Hence we obtain this expression

$$1xy + 0x(1-y) + 0(1-x)y + 0(1-x)(1-y). \tag{3.1}$$

As a second example, when we consider the disjunctive proposition, ie,. 'Either x is true, or Y is true,' and assume its members to be exclusive, then the addition sign denies the first case and the last. Therefore the cases to be considered are

2nd	X True, Y False	$x(1-y)$
3rd	X False, Y True	$(1-x)y$

The sum of these elective expressions equated to unity yields

$$x - 2xy + y = 1.$$

In expanding the function, as it has been done above, then we have this expression

$$0xy + 1x(1-y) + 1y(1-x) + 0(1-x)(1-y). \tag{3.2}$$

As a third example, let us consider the proposition, Either X is true or Y is true, and assume that the two members of the proposition are not exclusive. The addition sign denies the last case. Hence the cases to be considered are

1st	X True, Y True	xy
2nd	X True, Y False	$x(1-y)$
3rd	X False, Y True	$(1-x)y$

The sum of these elective expressions equated to unity gives this function

$$x + y - xy = 1$$

of which the expansion yields the following expression

$$1xy + 1x(1-y) + 1(1-x)y + 0(1-x)(1-y). \qquad (3.3)$$

Lastly we consider the expression of the conditional proposition, if X is true, then Y is true, here it is implied that all the cases of X being true, are cases of Y being true. It is expressed by

$$x(1-y) = 0.$$

Hence the second case is denied, and thus the cases to be considered are

1st	X True, Y True	xy
3rd	X False, Y True	$(1-x)y$
4th	X False, Y False	$(1-x)(1-y)$

The sum of these elective expressions equated to unity gives

$$xy - x + 1 = 1.$$

When we develop the function we have

$$1xy + 0x(1-y) + 1(1-x)y + 1(1-x)(1-y). \qquad (3.4)$$

As a result, it appears that Boole has the idea of all possible distribution of truth-values. As he puts it,

> It is evident that if the number of elective symbols is m, the number of the moduli will be 2^m, and that their separate values will be obtained by interchanging in every possible way the values 1 and 0 in the places of the elective symbols of the given function. (Boole 1847, p. 63)

Boole's Computational Procedure in Logic

Accordingly, Boole's procedure of development may be construed as the truth-table representation of a logical function, inasmuch as it pictures both those constituents corresponding to truth-possibilities, which the formula matches, as well as those which it does not. In this sense, we may extend the notion of 'truth-table' to correspond to expression such as,

$$1xy + 0x(1-y) + 0(1-x)y + 0(1-x)(1-y),$$

which is obtained by giving the values 1 and 0 to the arguments of the function xy, and thus to have as values 1, 0, 0, 0 for the constituents. We may lay aside here the meaning of these values, which is not to be taken into account in the procedure of computation. In the language of mathematics this table is a function from the set $\{1, 0\}$ into the set $\{1, 0, 0, 0\}$. Thus the conjunction of the two propositions X and Y yields the following truth-table:

X	Y	XY
1	1	1
1	0	0
0	1	0
0	0	0

The propositional interpretation of this truth-table would be the truth-functional account of the word 'and' which is true if and only if both X and Y are true.

Likewise, the constituents of the expression (3.2) have as values 0, 1, 1, 0, when we give the values 1 and 0 to the arguments of the function

$$x - 2xy + y.$$

Hence we have the truth-table:

X	Y	$X+Y$
1	1	0
1	0	1
0	1	1
0	0	0

The logical interpretation of this truth-table would be the truth-functional account of the words 'either...or' interpreted as an exclusive disjunctive proposition, which is true if and only if X is true and Y false or Y is true and X is false.

The constituents of the expression (3.3) have as values 1, 1, 1, 0, when we give the values 1 and 0 to the arguments of the function

$$x + y - xy.$$

Then we have the following truth-table:

X	Y	$X + Y$
1	1	1
1	0	1
0	1	1
0	0	0

The propositional sense of the truth-table would be the truth-functional account of the words 'either... or' interpreted as an inclusive disjunctive proposition, which is false if and only if both X and Y are false.

Finally, the constituents of the expression (3.4) have as values 1, 0, 1, 1, when we give the values 1 and 0 to the arguments of the function

$$xy - x + 1$$

Hence we have this truth-table:

X	Y	$X \Rightarrow Y$
1	1	1
1	0	0
0	1	1
0	0	0

which corresponds to the truth-functional account of the conditional proposition 'if... then', which is false if and only if X is true and Y is false. It is worth noting that this treatment of the conditional as true where the antecedent is false is congenial to the 'Boolean' treatment of universal propositions as lacking existential import.

Boole's anticipation of modern logic has been stressed by Kneale, who says that the chief novelty in Boole's system is his theory of truth-functions and their expression in disjunctive form (Kneale 1962, p. 420). He shows that Boole has all that is needed for an interpretation of his system in terms of the truth-values of propositions, and therefore he should not have dropped the promising suggestion laid down in *The Mathematical Analysis of Logic* (Kneale 1962, pp. 413–414).

Moreover, according to Kneale, it has been proved by P. Henle[7] that the most economical system of modal logic is equivalent to Boole-Schröder algebra (not the two-valued algebra) when the elements of the algebra are taken to be propositions and the modal sign $\Diamond$ is introduced by the rules:

$$\Diamond P = 1 \text{ if and only if } P \neq 0,$$
$$\Diamond P = 0 \text{ if and only if } P = 0.$$

Then Kneale claims that 'this system, with its strong reduction principle, is the working out of what Boole had in mind when he suggested in his *Mathematical Analysis of Logic* that the sign 1 might be taken to represent the sum of all possibilities, which he called the Universe of the Proposition' (Kneale 1962, p. 552).

Hailperin too holds that 'Boole need not have given up this approach which he had adopted in *Mathematical Analysis of Logic*, for when suitable clarifications and corrections are made, still remaining within the ambit of his ideas, a viable logic of propositions does result' (Hailperin 1984, p. 40). He builds up, in accordance with modern standards of rigour, a formal system for Boole's calculus of elective symbols.

Furthermore, whilst sharpening the point that Boole made great accomplishments in logic before Frege carried out his *Begriffsschrift*, Boolos corroborates as well the above interpretation of Boole's propositional calculus. He points out indeed that in the next-to-last section of *The Mathematical Analysis of Logic*, called 'Properties of Elective Functions,' Boole clearly had the idea of all possible distributions of truth-values. Thus, in his eyes, one main feature of the method of truth-tables, usually credited to Post and Wittgenstein, was on prominent display in Boole's early monograph (Boolos 1998, pp. 244–245).

It is said that a few years after he wrote it, Boole described the booklet as 'a hasty and regretted publication'. But Kneale reveals that 'towards the end of his life Boole said that he was dissatisfied with the exposition and arrangement of his *Laws of Thought*, and that he wished he had spent twice as long in working out the ideas first presented in his *Mathematical Analysis of Logic*' (Kneale 1947, p. 173).

3.5.2 Propositional Calculus and 'Time'

In *The Laws of Thought*, Boole adopts the notion of 'time for which a proposition is true,' and then bases the propositional calculus upon his calculus of classes. He suggests establishing a system of notation for the expression of

---

[7]See Lewis and Langford, *Symbolic Logic* p. 501.

secondary propositions, in which the symbols are exposed to the same laws of combination as the corresponding symbols employed in the expression of primary propositions. He writes:

> Let us employ the capital letters X, Y, Z, to denote the elementary propositions concerning which we desire to make some assertion touching their truth or falsehood, or among which we seek to express some relations in the form of a secondary proposition. And let us employ the corresponding small letters x, y, z, considered as expressive of mental operations, in the following sense, viz.: let x represent an act of the mind by which we fix our regard upon that portion of time for which the proposition x is true; and let this meaning be understood when it is asserted that x denotes the time for which the proposition X is true. Let us further employ the connecting $+$, $-$, $=$, &c., in the following sense, viz.: let $x + y$ denote the aggregate of those portions of time for which the propositions X and Y are respectively true, those times being entirely separated from each other. Similarly let $x - y$ denote that remainder of time which is left when we take away from the portion of time for which X is true, that by (supposition) included portion for which Y is true. Also, let $x = y$ denote that the time for which the proposition X is true, is identical with the time for which the proposition Y is true. We shall term x the representative symbol of the proposition X, &c. (Boole 1854, pp. 164–165)

Thus, Boole considers secondary propositions to concern the times at which primary propositions are true, and abstracts from them class terms by assembling into a class the times for which the propositional function is true. The small letters 'x', 'y', 'z', are expressed to stand for the time during which the propositions represented by the capital letters 'X', 'Y', 'Z', are true. From the above definition, it follows that

$$x + y = y + x,$$

for the two members of the equation denote the same sum of time.

Further, Boole expresses by 'xy' the whole operation of that portion of time in which the propositions X and Y are both true. From this definition we infer,

$$xy = yx.$$

For, as Boole sees it, whether we select, first that portion of time in which the proposition Y is true, then out of the result that contained portion in which

X is true; or first, that portion of time for which the proposition x is true, then out of the result that contained portion of it in which the proposition Y is true; we shall arrive at the same final result, that is, that portion of time in which the propositions X and Y are both true.

Then Boole adds to the above laws this one,

$$x(y + z) = xy + xz,$$

and the fundamental law of his logical system, that is,

$$x(1 - x) = 0.$$

There follows Boole's laws of secondary propositions, which are summed up as follows:

$$x + y = y + x$$
$$xy = yx$$
$$x(y + z) = xy + xz$$
$$x^2 = x \Rightarrow x(1 - x) = 0$$

Thus, the commutative law, the distributive law, and the fundamental law of duality apply in Boole's calculus of proposition. As the index law is satisfied by the symbols '0' and '1', then it is required to give them an interpretation in the logic of proposition. Hence, for Boole, in the expression of secondary propositions, '0 represents nothing in reference to the element of time', and '1 represents the universe, or whole time, to which the discourse is supposed in any manner to relate.' Hence, in secondary propositions, the universe of discourse may be a single day or a passing moment, or the whole duration of time, that is, the eternity.

Boole proceeds next to deduce general rules for the expression of secondary propositions. Since '1' expresses 'all time' and 'x' the time for which the proposition 'X' is true, so '$1 - x$' expresses the time for which the proposition 'X' is false. As 'xy' expresses the time for which the propositions 'X' and 'Y' are both true, when it is combined with the formula that expresses their falsehood, then we have the following interpretations:

$$x(1-y),$$

the time in which the proposition X is true and the proposition Y false;

$$(1-x)(1-y),$$

the time in which the proposition X and Y are simultaneously false;

$$x(1-y) + y(1-x),$$

the time in which either X is true or Y is true, but not both true; and

$$xy + (1-x)(1-y)$$

the time in which X and Y are either both true or both false. It should be added that the same principle may be applied when another symbol z occurs.

Subsequently, Boole can state the laws of the expression of propositions. Thus, the equation

$$x = 1$$

expresses the proposition x is true at all times. The expression

$$x = 0$$

means that the proposition X is false at all times. The equation

$$x(1-y) + y(1-x) = 1$$

expresses the disjunctive proposition, 'at all times either the proposition X is true or the proposition Y is true' with the assumption that the two propositions are mutually exclusive. If they are not exclusive, then we add to the left hand side of the above equation xy to have

$$x + (1-x)y = 1.$$

The conditional proposition, 'if the proposition Y is true, the proposition X is true,' is expressed as

$$y = vx.$$

Boole interprets this as: the time in which the proposition Y is true is some of the time in which the proposition X is true; that is, to say, that it is some indefinite portion of the whole time in which the proposition X is true. The symbol 'v' stands for time indefinite. But at this point Boole's approach is problematic. The expression 'vx' is normally used by Boole to express 'Some

Boole's Computational Procedure in Logic 91

X', and thus implies existence. In the present context, therefore, if Boole uses '$y = vx$' to express the conditional 'if Y then X', it will follows that the conditional as a whole is false if Y is false, contrary to the truth functional account of the conditional as the familiar 'material conditional' set out in *The Mathematical Analysis of Logic* (see subsection **3.5.1**). Boole however avoids this implication by allowing here that even if there is no time when X is true, still '$y = vx$'. This indeed preserves the treatment of the conditional as material; but it shows the problematic nature of Boole's use of 'v'.

Lastly, the following propositions

    1st.   If either X is true or Y is true, then Z is true
    2nd.   If X is true, then either Y is true or Z true
    3rd.   If either X is true or Y is true, then either Z and W are both true, or they are both false,

in which the conditional and the disjunctive both exist, are expressed respectively as,

$$x(1-y) + (1-x)y = vz$$
$$x = v(y(1-z) + z(1-y))$$
$$x(1-y) + y(1-x) = v(zw + (1-z)(1-w))$$

It should be observed that Boole does not represent 'X is sometimes true'. There is also no symbol for inclusion between classes, and therefore no symbol for the implication between propositions. In his calculus of classes, 'All Xs are Ys' or 'Xs are included in Ys' is represented by '$x = vy$'. Identically, 'All times when X is true are times when Y is true' or 'If X then Y' or 'X implies Y' is '$x = vy$'. Hence, Boole would conceive of the logic of propositions as incorporated in the Aristotelian syllogistic by reducing the form If 'X then Y' to 'All times when X is true are times when Y is true'.

It follows that Boole's procedures of computation, which have been so far accounted and illustrated in primary propositions, will be operational in the propositional calculus without any modification.

Boole did not clearly explain why he abandoned the direction taken in *The Mathematical Analysis of Logic* for interpreting the symbol 1 as the universe of 'cases'. In *The Laws of Thought*, he only tried to explain, in a vague way, what he considered as unsatisfactory in such an interpretation. He wrote,

> In a former treatise on this subject (*Mathematical Analysis of Logic*, p. 49), following the theory of Wallis respecting the Reduction of Hypothetical Propositions, I was led to interpret the symbol 1 in

secondary propositions as the universe of 'cases' or 'conjectures of circumstances;' but this view involves the necessity of a definition of what is meant by a 'case', or 'conjecture of circumstances;' and it is certain, that whatever is involved in the term beyond the notion of time is alien to the objects, and restrictive of the processes, of formal Logic. (Boole 1854, p. 176)

Boole hinted first at a philosophical reason why he interpreted the symbol 1, in *The Laws of Thought*, not as the universe of cases, but the eternity. He took it that secondary propositions involve the notion of time, to some extent in accordance with the view of Kant who regards time as one of the forms of the human understanding, and therefore one of the conditions of knowledge imposed by our own minds upon all that is submitted to its apprehension. Thus Boole thought that the formal processes of reasoning in secondary propositions require, as an essential condition, the occurrence in time of the propositions about which we reason. Hence he regarded the idea of time as essential to the establishment of a theory of secondary propositions.

However, Boole then dismissed these metaphysical speculations, and held that

The reason why the symbol 1 in secondary propositions represents, not the universe of events, but the eternity in whose successive moments and periods they are evolved, is, that the same sign of identity connecting the logical members of the corresponding equations implies, not that the events which those members represent are identical, but that the times of their occurrence are the same. These reasons appear to me to be decisive of the immediate question of interpretation. (Boole 1854, p. 176)

Indeed Boole interprets a hypothetical proposition such as 'If the sun shines, then the day will be fair' into: the class of moments of time in which the proposition 'the sun shines' is true is some of the class of moments of time in which the proposition 'the day will be fair' is true. This amounts to identifying the extension of the concepts which are specified as classes of moments of time in which a proposition is true. But, as Frege pointed out, 'this conception has the disadvantage that time becomes involved where it should remain completely out of the matter' (Frege 1882c, p. 93).

When McColl introduced implication as a connective symbol to treat secondary propositions, he separated Boole's secondary propositions from his primary propositions (see section **5.5**). Thus Frege observed that McColl succeeded in avoiding intermingling time and logic. But, as a result, 'every interconnection is severed between the two parts which, according to Boole, compose logic. We proceed, then, either in *primary propositions* and use the formulas in the sense stipulated by Boole; or else, we proceed in *secondary propositions* and use the interpretations of McColl' (Frege 1882c, p. 93). So the logical transition from primary propositions to secondary propositions, or the other way around, becomes impossible. However, as will be shown later, Frege set up a simple and appropriate organic relation between Boole's primary and secondary propositions, and provided their homogeneous presentation through his unified logical system (see subsection **6.4.1**).

3.6 Some Criticisms of Boole's Logical Calculus

Although logicians and historians of the discipline acknowledge the importance of Boole's logical calculus and give it an important place in the history of logic, it is nonetheless striking that many of them are not familiar with his work. Having explained the basic structure of Boole's theory, I shall here discuss some criticisms levelled at Boole.

3.6.1 Boole's 'Mathematicism'

The central charge laid against Boole's logical calculus may be encapsulated in what Bochenski calls 'Boole's mathematicism', which is his propensity to introduce mathematical procedures into the process of thinking, so that he fails to keep the logical meaning of the symbols. As he puts it,

> Boole's mathematicism goes so far... that he introduces symbols and procedures which admit of no logical interpretation, or only a complicated and scarcely interesting one. Thus we met with subtraction and division and number greater than 1. (Bochenski 1970, p. 298)

Boole would reply to such a criticism by comparing the introduction of uninterpretable symbols in the procedure of computation with the use of imaginary numbers in mathematics. As he says,

> The chain of demonstration conducting us through intermediate steps which are not interpretable, to a final result which is interpretable, seems not only to establish the validity of the particular application, but to make known to us the general law manifested therein.... The employment of the uninterpretable symbol $\sqrt{-1}$, in the intermediate processes of trigonometry, furnishes an illustration of what has been said. (Boole 1854, p. 69)

The general principle underlying this procedure is that the formal processes of reasoning depend only upon the laws of the symbols, and not upon the nature of their interpretation.

But is it tolerable that Boole introduces uninterpretable expressions in his logical calculus, even if they are eliminated by a process of interpretation? Feys answers this question as follows,

> We have not to do with any nonsense; some steps would make nonsense in modern class algebra, but every step of the method makes sense in ordinary algebra, as a means of obtaining special forms of solution in it-let us say solutions such that $xx = x$ and with no $\frac{1}{0}$. In fact the method may be compared with the biblical method in dealing with cockle: to let it grow together with wheat, then to separate and to burn the cockle away. (Feys 1955, p. 100)

Since the cockle is not a parasite which is liable to damage the wheat, the method works well enough. Likewise, Boole's procedure relies provisionally upon uninterpretable symbols in the computation, but eventually they are expunged by the interpretation. The procedure may be described as follows: Boole expresses a problem by means of an equation in which symbols have a logical meaning; he then carries out operations on them as though the symbols are restricted to the values 0 and 1, giving them a mathematical meaning, then he restores them to their logical meaning by using the development and interpretation procedure. The final result is both valid and interpretable; and the difficulties arising from the uninterpretable symbols are thereby circumvented. In truth the final result obtained by the procedure does not depend upon the meaning of the symbols but only upon their formal laws.

Yet by the time of the non-symbolic Aristotelian logic, the assumption was that throughout a valid reasoning words must hold the same meaning - and a meaning known previously. But it turns out that modern formalised reasoning is not concerned at all with the meaning of the symbols. As Carnap defines it,

> A theory, a rule, a definition, or the like is to be called *formal* when no reference is made in it either to the meaning of the symbols (for example, the words) or to the sense of the expressions (e.g. the sentences), but simply and solely to the kinds and order of the symbols from which the expressions are constructed. (Carnap 1934, p. 1)

In his book, *The Logical Syntax of Language*, Carnap maintains the view that we have in every respect complete liberty with regard to the forms of language; both the forms of construction for sentences and the rules of transformation

Boole's Computational Procedure in Logic 95

may be chosen quite arbitrarily. He calls this standpoint the 'principle of tolerance', that is, 'it is not our business to set up prohibitions, but to arrive at conventions' (Carnap 1934, p. 51). Hence, 'in logic, there are no morals'; 'everyone is at liberty to build up his own logic, ie. his own language, as he wishes' (Carnap 1934, p. 52).

In keeping with the above, in the section on 'Formalism' in his book, *Quiddities*, Quine defines formalism as disinterpretation; i.e. formulating purely in terms of syntactic manipulation. He goes on to say:

> Today formalism is the stock in trade of the thousands of nononsense technicians who make their living programming computers. A computer requires blow-by-blow instruction, strictly in terms of what to do to strings of marks or digits, and eked out by no armwaving or appeals to common sense and imagination; and such, precisely, is formalism. (Quine 1987, p. 67)

It turns out that Boole, without going as far as this, has a clear idea that symbols may be employed, without depending upon a determinate meaning, and such an idea seems to dovetail with the above modern definitions of formalism. As Putnam sees it, 'in fact, Boole was quite conscious of the idea of *disinterpretation*, of the idea of using a mathematical system as an algorithm, transforming the signs purely mechanically without any reliance on meanings' (Putnam 1992, p. 254).

C. I. Lewis, in his *A Survey of Symbolic Logic*, notices that 'Boole allows operations which have no direct logical interpretation, and is obviously more at home in mathematics than in logic'; he acknowledges that

> It is probably the great advantage of Boole's work that he either neglected or was ignorant of those refinements of logical theory which hampered his predecessors. The precise mathematical development of logic needed to make its own conventions and interpretations; and this could not be done without sweeping aside the accumulated traditions of the non-symbolic Aristotelian logic. (Lewis 1960, p. 51)

Moreover, criticisms that Boole introduces logically uninterpretable expressions in his logical calculus are misguided in that they rest upon a misunderstanding of the postulate of his general method in logic, that is, the manipulation of an independent formal procedure of computation viewed as a support of logical deductions. Boole considers most of these uninterpretable expressions as belonging to the procedures of computation which allow him to carry out the requisite formal processes in order to work out logical problems. They are not logical expressions, and Boole has never claimed the contrary. This can be substantiated by a close analysis of *The Laws of Thought*.

In this book Boole separates sharply his fundamental logical system from the procedures of computation. They are carried out in different chapters: Chapter II presents the fundamental logical system, upon which depend Chapters V–VIII in which Boole's computational procedures occur. The so-called logically uninterpretable expressions appear in these chapters where Boole carries out his 'general method in logic.' He specifies the difference between these two aspects of his work as follows:

> The previous chapters of this work have been devoted to the investigation of the fundamental laws of the operations of the mind in reasoning; of their development in the laws of symbols of logic; and of the principles of expression, by which that species of propositions called primary may be represented in the language of symbols. These inquiries have been in the strictest sense preliminary. They form an indispensable introduction to one of the chief objects of this treatise-the construction of a system or method of logic upon the basis of an exact summary of the fundamental laws of thought. (Boole 1854, p. 66)

Thus, Boole considers his fundamental logical system to serve as a necessary prelude to his general method. It is confined to manipulating interpreted logical operations, whereas the general method applies computational procedures, which are not exclusively logical. Boole describes his general method while wondering '... whether it is necessary to restrict the application of symbolical laws and processes by the same conditions of interpretability under which the knowledge of them was obtained.' According to Boole, 'if such restriction is necessary, it is manifest that no such thing as a general method in logic is possible.' But 'on the other hand, if such restriction is unnecessary, in what light are we to contemplate processes which appear to be uninterpretable in that sphere of thought which they are designed to aid?' (Boole 1854, p. 66) As might be expected, Boole claims such a restriction to be unnecessary and then answers that question.

He observes that this apparent failure of correspondence between process and interpretation does not manifest itself in the ordinary applications of human reason. 'For no operations are there performed of which the meaning and the application are not seen'. Accordingly, there are many who would be disposed to extend the same principle to the general use of symbolical language as an instrument of reasoning.

Boole's Computational Procedure in Logic

They may argue that

> As the laws or axioms which govern the use of symbols are established upon an investigation of those cases only in which interpretation is possible, we have no right to extend their application to other cases in which interpretation is impossible or doubtful, even though (as should be admitted) such application is employed in the intermediate steps of demonstration only. (Boole 1854, p. 67)

But, in Boole's eyes, this objection itself is fallacious. For, as he puts it,

> Whatever our *à priori* anticipations might be, it is an unquestionable fact that the validity of a conclusion arrived at by any symbolical process of reasoning, does not depend upon our ability to interpret the formal results which have presented themselves in the different stages of the investigation. (Boole 1854, pp. 67–68)

We have here the foundation of the general method in logic which is stated in this postulate:

> We may in fact lay aside the logical interpretation of the symbols in the given equation; convert them into quantitative symbols susceptible only of the values 1 and 0 ; perform upon them as such all the requisite processes of solution; and finally restore them to their logical interpretation. (Boole 1854, p. 69)

It is clear that this procedure is different from the one at work in the fundamental logical system of Chapter II. It may be exemplified by Boole's process of development of any functions, logical or not, into constituents, which does not need to bring out logically interpretable expressions. Let us illustrate these two cases in which occur the so-called uninterpretable expressions.

The first case is apparent in Boole's definition of men, viz., 'men are rational animals', which is expressed as

$$x = yz.$$

With x referring to the set of 'men', y to the set of 'rational beings' and z to the set of 'animals', Boole works out a definition of 'rational beings' in terms of 'men' and obtains the equation

$$y = \frac{x}{z}.$$

Since, as a division, the right hand side of the equation is not logically interpretable, he then develops it and obtains,

$$y = 1zx + 0z(1-x) + \frac{0}{0}(1-z)(1-x) + \frac{1}{0}(1-z)x.$$

According to Boole's canons of interpretation this expression can be rewritten as

$$y = zx + v(1-z)(1-x)$$
$$(1-z)x = 0$$

which are respectively interpreted into

> 1st. Rational beings consist of all animals that are men (...) and an indefinite remainder (some none or all) of beings that are neither animals nor men.
>
> 2nd. Men that are not animals do not exist. (Boole 1856, p. 98)

Boole makes an observation upon this symbolical equation in which occur what have been criticised as uninterpretable expressions. In his eyes, 'the equation exhibits the result of the abstraction as obtained by the process of development that process depending not upon the meaning of the symbols z, x, but only upon their formal laws' (Boole 1856, p. 99). Thus, it is clear that while processing the expansion of the function

$$y = \frac{x}{z},$$

Boole does not consider the symbols such as '$\frac{1}{0}$' and '$\frac{0}{0}$' involved in the process as logical at all. They are the result of the formal treatment of z, x, as quantitative symbol restricted to the values 0 and 1. But, for him, in the resulting expression, the coefficient '$\frac{1}{0}$' attached to the constituent '$(1-z)x$' indicates not only that no part of the class represented by that constituent is included but that the existence of such class is forbidden by the premises. The coefficient '$\frac{0}{0}$' attached to the constituent '$(1-z)(1-x)$' implies that an indefinite portion of the class represented by that concept is signified (Boole 1856, p. 99).

The second case in which the charged uninterpretable symbols occur can be found in the first example of development that Boole himself gives. He expands the function

$$\frac{1+x}{1+2x},$$

and obtains,

$$\frac{1+x}{1+2x} = \frac{2}{3}x + 1 - x.$$

This is an example of development which is not carried out in a logical context, and Boole does not claim that this formula is logically interpretable. He even specifies that 'by the principle which has been asserted in this chapter, it is lawful to treat x as a quantitative symbol susceptible only of the values 0 and 1' (Boole 1854, p. 72).

It follows that the criticisms of the symbols such as, '$\frac{1}{0}$' and '$\frac{0}{0}$' are misguided. Indeed, these uninterpretable logical symbols mostly occur as coefficients of the constituents of an expanded function. The critics claim that they are both logical and uninterpretable. However, in Boole's work, they are either non-logical expressions which appear in the intermediate steps of the process of deduction which yields a logically interpretable result according to the rule of the general method, or they belong to expressions which constitute an illustration of non-logical examples of the procedure of development. In both these cases in which the uninterpretable expressions in question occur, it cannot be said with cogency that Boole's position is illegitimate.

Regarding now the complaint that Boole introduces division into logic, at first sight this seems irrelevant since in Chapter II of *The Laws of Thought* he clearly does not count it as a logical operator. He claims

> ...the axiom of algebraists, that both sides of an equation may be divided by the same quantity, has no formal equivalent here. (Boole 1854, pp. 36–37)

Thus, Boole refuses to consider division as a logical operator. However, it may be argued that he introduces division in his logical calculus via the expression '$\frac{0}{0}$'. For he says,

> ...as in Arithmetic, the symbol $\frac{0}{0}$ represents an *indefinite* number, except when otherwise determined by some special circumstance, analogy would suggest that in the system of this work the same symbol should represent an *indefinite class*. (Boole 1854, p. 89)

Hence it follows that the expression '$\frac{0}{0}$', which is supposed to be a mathematical expression belonging to his procedures of computation, is counted as an elective symbol such as, x, y, z of his fundamental logical system. Hence, Boole contradicts himself since he claims to exclude division in his logical calculus. But, in Van Evra's eyes, it is not clear that Boole intends '$\frac{0}{0}$' to function as an elective symbol of his fundamental logical system. On the contrary, he argues that 'Boole seems to use $\frac{0}{0}$ not to stand directly for an indefinite class, but as a way of indicating the irrelevance of class size, in certain cases, for the validity of inferences in which they occur' (Van Evra 1977, p. 370). Thus, the expression would then be restricted to the general method, and not be a part of Boole's fundamental logical system.

Yet it should be noted that Peirce provides a means that enables one to eliminate division in Boole's logical calculus. For him, the rule for clearing fractions is the definition of division, which he gives as follows,

> Let $x = \frac{b}{a}$. Then the question we have now to ask is what is the meaning of x in terms of b and a. The answer is $ax = b$. That is x is a class which when determined by the comprehension of a gives b. (Peirce 1865, p. 227)

Thus, $\frac{b}{a}$ is a class which, when multiplied by a yields b, that is, it comprises all b and some, all, or none of what is not a beside. So, the inverse process of division implies that the comprehension of the dividend includes the comprehension of the divisor. It follows that to find what $\frac{0}{0}$ means, we put it equal to x, $\frac{0}{0} = x$; multiply by 0 and then obtain $0 \times x = 0$. Now this is true whatever be the value of x. Hence if $\frac{0}{0} x = y$, then $0 \times x = 0 \times y$. But this is true whatever the relationship between x and y—whether y is *all* of x, *some* of x or *none* of x. As for the meaning of $\frac{1}{0}$, we must remember that when one letter is divided by another, it must be that the dividend contains the divisor as a factor. Now to say that 1 contains 0 as a factor amounts to saying that it enters into the meaning of nothing and does not exist. This same result may also be obtained when we let $\frac{1}{0} = y$; then multiplying by zero, we have $0 \times y = 1$, which means that the class common to y and to nothing is the universal class. But no such class exists or is conceivable.

In the case of subtraction the issue is that subtraction can have a logical meaning only at the condition of giving an exclusive interpretation to addition. This works in arithmetic, but is not suitable for logical sum. Jevons' introduction of a non-exclusive disjunction, however, shows how Boole's treatment of disjunctive as exclusive can be set aside. It enables Jevons to formulate the self-evident law of thought, the law of unity, '$x + x = x$', which he considers as inconsistent with Boole's logical calculus (Jevons 1864, p. 74). He claims that since Boole assumes as a condition of his system that each two terms must be mutually exclusive, he cannot recognise the law of unity. Indeed, in Boole's logical calculus the expression

$$x + (1 - x)$$

denotes all xs with all not-xs, which taken together, yield all things, or 1 the 'universe of discourse'. As Jevons suggests, let us attempt, by multiplication with x, to select all xs from this expression for all things,

$$x(x + 1 - x) = x + x - x.$$

Boole's Computational Procedure in Logic   101

Boole would obtain x as the required expression for all xs by deleting $+x$ and $-x$. However, for Jevons, it is self-evident that

$$x + x = x.$$

Hence
$$x + x - x = 0$$

and not
$$x.$$

For Jevons, who interprets non-exclusively the symbol '+', the above expression is 'nonsense' and not null set as the '0' which means 'excluded from thought' indicates. It follows that Boole's operation of subtraction is inconsistent with Jevons' self-evident law of unity, which eliminates expressions such as '$2x$' that have no interpretation.

3.6.2 The Inconsistent Treatment of the Symbol 'v'

Beyond any doubt, it should be admitted that, in *The Laws of Thought*, the way in which Boole handles the symbol 'v' appears to be very confusing. He considers the affirmative universal proposition, 'All men are mortal' as meaning that 'All men are some mortal beings', and employs the symbol 'v' in order to express 'some'. He expresses the proposition as follows,

$$y = vx.$$

Thus, it is assumed that the class which the symbol 'v' represents must not be empty and that it is possible that this class is the universal class. The symbol 'v' is handled here, as though it were a class symbol. As Boole stresses,

> It is obvious that v is a symbol of the same kind as x, y, &c., and that it is subject to the general law, $v^2 = v$, or $v(1-v) = 0$. (Boole 1854, p. 61)

So Boole conceives of 'v' as representing a non-empty class of indefinite extension which may even be the universal class. In truth this seems to be a mistake since there is no such class, and what is indefinite is not the class represented by 'v' but whether 'v' represents a class of one extension or another. That is, it is to be indefinite whether 'v' represents one non-empty class or another-any will do, including the universal class.

Jevons believes that Boole's way of expressing the affirmative universal proposition by using the symbol 'v' only introduces complexity, and destroys the beauty and simple universality of the system which may be created without its use. Accordingly, in *The Principles of Science*, Jevons declares: 'throughout this system of logic I shall dispense with such indefinite expressions.... (Jevons 1874, p. 41) He does so by expressing the proposition 'All As are some Bs' in the form: '$A = AB$'. This form was probably first employed by Leibniz in his *Difficultates quædam logicæ* saying '*omne A est B; id est equivalent AB et A*'.[8]

Moreover, Boole will suggest later the elimination of the symbol from the expression of universal propositions. For instance, he regards the affirmative universal proposition,

$$y = vx,$$

as having existential import, inasmuch as the symbol v is not totally indefinite and is fixed so as to have 'some of its members... mortal beings.' Thus, vx denotes 'some x' or, 'x is not empty'. But, this device for the expression of existential propositions is in practice eliminated by Boole with his procedure of elimination which consists of substituting 1 and 0 for v in

$$y - vx = 0,$$

and of multiplying the resulting equations together to have,

$$y(1 - x) = 0.$$

As a result, the existential import of the universal affirmative is discarded.

However, regarding the particular affirmative 'some xs are ys', which is expressed as

$$vx = vy,$$

Boole introduces 'v as the symbol of a class indefinite in all respects but this, that it contains some individuals of the class to whose expression it is prefixed' (Boole 1854, p. 63). Thus, since the meaning of v depends on what it is prefixed to, it is not the same as a class symbol. So, further, he says '... v is not quite arbitrary, and therefore must not be eliminated. For v is the representative of *some*, which, though it may include in its meaning *all*, does not include *none*' (Boole 1854, p. 124).

Hence Boole gives an inconsistent treatment to the symbol 'v'. He treats the symbol 'v' in the universal affirmative

$$y = vx$$

---

[8] See Venn 1881, p. 176

differently from the one in the particular affirmative

$$vx = vy.$$

He discards the existential import of 'v' from the first expression, but leaves it as it stands in the second one. It follows that Boole's use of the indefinite class symbol 'v' is inconsistent with respect to existential import.

In addition, Boole expresses the particular negative 'some xs are not ys'

$$vx = v(1-y),$$

where v stands for the indefinite class *some*. But as Peirce sees it, the absurdity of this is evident from the fact that by transposing we get

$$vy = v(1-x)$$

or 'some ys are not xs'. But it does not follow from 'some xs are not ys' that 'some ys are not xs'. This expression is therefore wrong. Indeed, in his paper, "Description of a Notation for the Logic of Relatives" (1870, pp. 422–425), Peirce criticises Boole's treatment of particular propositions in detail.

A defect of rigour which occurs in Boole's logical calculus while handling the symbol 'v' can be also shown in this example: from the equations

$$vx = vy$$

expressing 'some men are married', and

$$vx = v(1-y)$$

'some men are unmarried', it is obtained by substitution

$$vy = v(1-y)$$

expressing 'some married are unmarried' which is contradictory.

Another misunderstanding that the symbol 'v' leads to is that Boole interprets the expression $\frac{0}{0}$ as indicating that *all*, *some*, or *none* of the class to whose expression it is affixed must be taken, and claims that it may be replaced by an uncompounded symbol v (Boole 1854, p. 90). Thus, 'v' is understood to include *all*, *some*, *none*. However, he holds, as well, that 'v is the representative of *some*, which, though it may be include in its meaning *all*, does not include *none*' (Boole 1854, p. 124). This is also a defect of rigour that incurs in Boole's work. As Venn notes, if we put 'v' as an equivalent for $\frac{0}{0}$, this 'v' is by no means a substitute for the 'some' of ordinary logic, since it includes 'nothing'; still less for the 'some' of ordinary language, since it also includes 'all'.

Therefore, Venn says,

> It is the more necessary to call attention to this since Boole himself has repeatedly treated this v as equivalent to 'some', which I can only regard as an oversight. (Venn 1881, 175. See also p. 161n.)

Undoubtedly, the symbol 'v' is a baffling problem in Boole's logical calculus. In fact it is not a suitable way of expressing particular existential judgments. But attempts were made by Peirce, Schröder and Whitehead to work out a way of repairing the flaw of Boole's 'v'. In effect, in his 1870 memoir, 'Description of the Notation for the Logic of Relatives', Peirce noticed that Boole's insistence on using equational forms prevented him from negating propositions and forced him to represent particular propositions with his symbol 'v'. However, according to Peirce, the simplest way to express particular propositions was to allow propositional negation and also inclusion (Peirce 1870, p. 423). So Peirce introduced a new symbol '$<$' and let '$A < B$' mean the same as 'A is included in B' (or 'All A are B') and it is not true that B is included in A (or it is not true that 'All B are A'). It follows that 'some animals are horses' can be expressed by $0 < a, h$ (or equivalently, $a, h > 0$), where the comma represents Boolean multiplication, and 0 stands for the null class. The traditional syllogistic forms can be shown valid in this symbolism. Thus Peirce's new symbol provided a way to express particular propositions. But the use of the symbol required him to drop the fundamental principle that all propositional forms must be represented in terms of equations or inclusions, for the definition of '$<$' involves the denial of an inclusion.

In his *Vorlesungen über die Algebra der Logik* (vol. ii 1891), Schröder too devised a simple expedient of expressing particular propositions. According to Burris (1998, pp. 31–33), Schröder pointed out that one could not use equation to handle existential statements. He then introduced negated equations by using the symbol $\neq$. The use of this notation allows one to have the correct translation of the traditional syllogistic forms into the language of the calculus of classes.

Regarding Whitehead's way of handling existential expressions, let me merely say that, in his *Universal Algebra* (1898, Bk. II, Chap. 3, pp. 81–97), he introduced new symbols, such as 'j' and 'ω', and let any combination of symbols involving them be called an existential expression. Thus 'xj' expresses that 'x' exists and '$x + \omega$' expresses that 'x' is not equivalent to the universe 'i' (Whitehead 1898, pp. 83–84). However, the complicated form of Whitehead's approach led Quine to wonder why he did not favour the simple way of expressing existence statements by means of inequalities: '$x \neq 0$' for 'Something belongs to x', and '$x \neq 1$' for 'Something does not belong to x'.

Quine's answer was that perhaps Whitehead was viewing algebras strictly as systems of equations as opposed to inequalities (Quine 1966, p. 10).

The flaw of Boole's 'v' seems to be repaired by writing particulars as inequations. As a result, a tidy system of calculus of classes is obtained which is easily decidable, and in some respect much more powerful than syllogistic (see section **4.2**). But it is the device of quantification introduced by Frege and Peirce which will set Boole right over particulars (see subsection **5.3.3** and section **7.1**).

However, despite the fact that Boole could not provide a suitable way of expressing particular existential judgements and that his computational procedure is a cumbersome piece of machinery with some defects of rigour, it remains, nonetheless, that in some respects his research programme went beyond Aristotle's. In effect, Boole came up with a mathematical technique which provides an algebraic treatment of the syllogism. Before Boole, the premises occurring in syllogistic theory involved only two terms. But when Boole introduced his new notation, inferences could be treated in a way which permitted any number of terms, and where the premises and conclusions involved any number of addition, multiplication and negation signs. Thus Boole performed an extension and improvement of Aristotelian logic. Furthermore, Boole's mathematical technique could express both syllogisms and propositional logic which Aristotelian logic did not cover. Boole's new notation could be used to represent a logical structure which might have an interpretation either in Aristotelian or in propositional logic. So Boole developed a truth-functional analysis of propositional logic whose formula is expressed in a disjunctive normal form. This important innovation goes beyond anything to be found in Aristotelian logic. Moreover, Aristotelian logic did not contain a formulation of the logic of relations, but de Morgan (1860), Peirce (1870) and Schröder (1895) whilst working within Boole's research programme developed a comprehensive theory of relations. This development was made possible through the systematic use of Boole's new mathematical technique and notation. In addition, in the course of the twenty century, Boolean logic was tidied up and presented in a strictly axiomatic form without reference to any special interpretation. One interpretation of the system was what Huntington called the algebra of logic which both provides a model for syllogistic but also for further matter and also provides a mathematical decision procedure. This refinement of Boolean algebra is even important for modern mathematicians.

Chapter 4

Some Further Developments of Boole's Logical Calculus

> *The time must come when the inevitable results of the admirable investigations of the late Dr. Boole must be recognised at their true value.*
>
> Jevons

Introduction

At the end of the nineteenth century, after Boole carried out the leading principles of the research programme for the 'Algebraic School', the task then fell to Jevons, Venn, Peirce, Schröder and Whitehead of developing systematically Boole's logical calculus. Thereafter, at the beginning of the twenty century, Sheffer and Huntington brought out the concept of 'Boolean algebra' whilst building up several sets of independent postulates for the algebra of logic, which was thereby presented as an axiomatic system.

In this chapter, I shall not be exploring every way in which Boole's logical calculus was developed. Rather, I shall be focussing mainly on the attempts of Venn to complete Boole's logical calculus. Then I shall describe, in the light of our modern interpretation, the algebraic expression of the calculus whose abstract formulation is known as Boolean algebra and its axiomatisation, which was set up by Sheffer and Huntington. In doing so, I want to point out the persistent recurrence of the same set of logical problems which Boole's research programme brought forth.

4.1 Venn: Diagrammatic Representation of Boole's Logical Calculus

Venn follows basically the way paved by *Mathematical Analysis of Logic* in attempting to establish Boole's formal procedures upon purely logical principles. He aims to give a logical interpretation of Boole's system, which would then be 'independent of the mathematical calculus.' But he does not jettison what he considers as being the 'most characteristic and attractive in Boole's system': the mathematical dress. Rather, he remains close to Boole, for it is important to be acquainted with mathematical formulae when dealing with logic. He then carries out, in symbolic logic, an extension of the signification of logical symbols similar to that existing in mathematics. Moreover, Venn regards his logical system as a calculus of classes in extension. In the *Symbolic Logic* published in 1881, he devises a diagrammatic method, which visualises the structure of Boole's logical calculus with such pictorial clarity that it becomes obvious to 'see' what this logic is all about.

I shall be concerned with Venn's systematic interpretation of propositions in term of existence, and the practical possibility it yields, namely the geometrical illustration of the operations between classes or the truth-values of propositions by means of diagrams.

4.1.1 A Compartmental View

The resort to diagrams so as to picture the structure of Boole's logical calculus presupposes an interpretation of propositions in terms of their existential import. In Venn's eyes, this is given through what he calls 'the compartmental view', which is based on the question of the occupation or non occupation of compartments. What is then required is a construction of notation for all possible combinations which any number of class terms may involve, and a mode of symbolic expression pointing out which of these distinct compartments are empty or occupied, according to the premises of the propositions in question. However, it turns out that Boole's logical symbolism is such a symbolic language, which spares Venn from building up another one.

The compartmental view is an extensional calculus of classes, which leads to a complete arrangement of all the compartments obtained by putting any number of classes together, and indicating whether objects exist that have the particular combination of attributes in question. Venn takes two class terms x and y, and considers the simple cases given by their combination. In this case, there are four possible compartments, because everything which exists must either have both the attributes expressed by x and y, or neither of them, or

Some Further Developments of Boole's Logical Calculus 109

one and not the other. With $\bar{x}$ standing for not-x and $\bar{y}$ for not-y, these four compartments are represented by this scheme:

$$xy = 0, \text{ or No } x \text{ is } y,$$
$$\bar{x}y = 0, \text{ or All } y \text{ is } x,$$
$$x\bar{y} = 0, \text{ or All } x \text{ is } y,$$
$$\bar{x}\bar{y} = 0, \text{ or Everything is either } x \text{ or } y.$$

xy stands for the compartment, or class of things which are both x and y, and the equation $xy = 0$ indicates that the compartment is unoccupied. $\bar{x}\bar{y}$ stands for the compartment, or class of things which are both not-x and not-y, and the equation $\bar{x}\bar{y} = 0$ indicates that the compartment is unoccupied. And likewise with the other equations.

Venn's approach brings up the problem of the import of propositions, as regards the existence of their subjects and predicates. In order to work out this problem, Venn raises two fundamental questions: whether the universal proposition 'All x is y' affirms or implies that there are such things as x or y? Or again does the particular proposition, 'Some x is y', make any different implication as to this special point of there being any x or y?

Regarding the particular proposition, 'Some x is y,' there is no difficulty: the existence of such things as x or y is required for the truth of this proposition; they have an existential import. It will be shown later that, in the diagrammatic method, a star is put in the conjunct of the two compartments for indicating that the conjunct is not unoccupied. Thus, the particular proposition may be read, 'Some, and there are some...'. It always implies the existence of members of the subject class.

What, however, of the universal proposition? If we say 'All x is y', does the proposition imply that there exists an x? According to Venn, a proposition asserting the existence of some members of the subject class does not need to be associated with the universal form. This is because he decides to express universal propositions systematically in negative terms. He claims,

> Every universal proposition may of course be put into a negative form: this is familiar to every logician, the distinction of our system being that this negative side of a proposition is more consistently and uniformly developed, and provided with a suitable symbolic notation. Now if we adopt the simple explanation (of a universal proposition) that *the burden of implication of existence is shifted from the affirmative to the negative form*; that is, that it is not the existence of the subject or the predicate (in affirmation) which is implied, but the non-existence of any subject which does not pos-

sess the predicate, we shall find that nearly all difficulty vanishes. (Venn 1881, p. 157-58)

This interpretation may be associated with a representation of classes as 'compartments,' to which the propositions indicate precisely whether they are occupied or unoccupied. Since particular propositions have an existential import, universal propositions should be analysed as the negation of the contradictory particular propositions, so as to eschew the difficulty stemming from the fact that the universal proposition does not have existential import. Under this negative form, there does not remain any uncertainty related to the interpretation of a universal proposition. For instance, the proposition 'All x is y' does not imply that the compartment xy is occupied, it shows that the compartment $x\bar{y}$ is unoccupied: it then shows that one of the four possible classes is empty. As Venn puts it,

> Instead of regarding the affirmative form as being the appropriate and unambiguous form, we regard the corresponding or equivalent negative form as possessing these attributes. Whether there be any xs or ys we do not know for certain, but we do feel quite sure that there is no such thing existing as 'x which is not y'. (Venn 1881, p. 158)

Venn interprets negatively the proposition 'all x is y', as saying, 'there are no such things as $x\bar{y}$', and expresses symbolically this proposition in the form $x\bar{y} = 0$. As he suggests it,

> Put the propositions 'All x is y', 'No x is y', into the forms $x\bar{y} = 0$, $xy = 0$, and the statements 'There is no $x\bar{y}$', 'There is no xy,' must admit of verification and be intelligible.... (Venn 1881, p. 144)

The advantage of this interpretation of the universal as existential negative is obvious when considering the combination of two propositions, as in the above scheme. Thus, the proposition 'All x is y,' is better written, 'No x is not-y' or $x\bar{y} = 0$, which shows that the class of $x\bar{y}$ is empty. But, it does not tell us anything about the three others alternatives. The symbols being subject to the universal condition expressed by

$$1 = xy + x\bar{y} + \bar{x}y + \overline{xy},$$

which means that the universe is the union of xy, x not-y, y not-x, not-x not-y, when one of these classes is empty, there are three classes left out. However, it is required that one, or two, or all three must be non-empty[1], but it is not

---

[1] Like Huntington's Postulate **VI** and Sheffer's Postulate **1** (see subsections **4.3.1** and **4.3.2**), Venn requires that the universe not be empty so as to keep the system from being vacuous.

clear which of these is non-empty. If now the first proposition, 'All x is y', is associated with a second that asserts that 'All y is x', then one of the three classes, $\bar{x}y$, is empty and there remains only two possible classes: xy or $\bar{x}\bar{y}$. A third proposition will then be required in order to assert that one only of these two classes is non-empty. If for example this third proposition asserts that 'xy is all', i.e. that the universe is both x and y, then the class $\bar{x}\bar{y}$ is empty, and xy is the only class which is non-empty.

It follows that the traditional square of opposition, which has been already tabulated (see subsection **2.1.3**), requires changing. For, the difference between the symbolic logic and traditional logic depends on the existence of members of the subject class. In effect, in traditional logic the subject of a universal proposition is assumed implicitly to have an existential import in the sense that the class denoted by the subject term is not empty. On the other hand, in symbolic logic the universal propositions do not have an existential import in the sense of not implying the existence of members of the class denoted by the subject term. Thus, even though Boole represents the universal proposition in the form $x = vy$ and implies for the subject of the proposition an existential import, while working out logical equations he always eliminates the 'v' obtaining $x(1-y) = 0$. The universal proposition represented in this form is non-existential since 'v' has dropped out. Hence, as in modern class algebra, Boole interprets 'All x is y' as being without existential import.

Venn, following Boole in this interpretation of 'All x is y', makes the problem of existential import more explicit. Indeed, according to the compartmental view, all the valid relations between the four propositions on the square of opposition are also valid, if there are members of the subject class. But only the relation of contradiction is valid if there are no members of the subject class. Thus the symbolical expressions of the relations between the four forms of propositions are: $x\bar{y} = 0$ and $x\bar{y} \neq 0$ for respectively the universal affirmative and the particular negative propositions which are contradictories; $xy = 0$ and $xy \neq 0$ for the universal negative and the particular affirmative propositions which are also contradictories.

Then what is called the Boolean square of opposition, which changes the appearance of the traditional square of opposition, may be tabulated as follows:

Figure 4.1

It can be noticed that contraries, subcontraries, and subalterns are missing in the Boolean square of opposition. For if the subject class of the universal proposition is empty, then the rule of the traditional square of opposition that **A** and **E** propositions, being contraries, cannot both be true does not hold. The rule that both subcontraries can be true but both cannot be false again does not hold if the subject class is empty. If an **A** proposition is true when the subject class is empty, its contradictory proposition is false. The contradictory is an **O** proposition, but because there are no members of the subject class, the corresponding **E** proposition is true. In this case, an **E** proposition is true when its subaltern **O** is false. This does not comply with the rule of the traditional square of opposition. A similar situation occurs between an **A** and its subaltern **I** when the subject class is empty.

4.1.2 A Geometrical Illustration of Boole's Logical Calculus

The compartmental view leads to the possibility of visualising the logical operations by means of diagrams drawn in such a way that each compartment of the picture represents one of the cases which the premise of a logical problem requires one to consider. Venn specifies the novelty of the method:

> I tried at first, as others have done, to illustrate the generalised processes of the symbolic logic by aid of the familiar method, but soon this was quite unsuitable for the purpose. Though the method here described may be said to be founded on Boole's system of logic,

Some Further Developments of Boole's Logical Calculus 113

> I may remark that it is not in any way directly derived from him. He does not make employment of diagrams himself, nor does he give any suggestions for their introduction. (Venn 1881, p. 114)

Before Venn, Euler[2] a Swiss mathematician of the eighteenth century, carried out a way of illustrating syllogistic inferences by using circles. But Venn comes up with an ingenious modification on Euler's circles by using two ideas which are fundamental in Boole's algebra of logic, the null class i.e. the class with no members and the universal class i.e. everything. Thus, Venn's diagrams are dissimilar to those of Euler in that Venn first pictures all possible combinations by distinct compartments, and then indicates, by marks within the various compartments, which combinations must be empty and which not empty according to the premise of a given logical problem. Venn describes what he intends to do in the following manner:

> What we propose to do is to form a framework of geometrical figures which shall correspond to the table of combinations of x, y, z, &c. All that is necessary for this purpose is to describe a series of closed figures, of any kind, so that each in succession shall intersect all the compartments already produced, and thus double their number. That this successive duplication is what is done with the letter symbols is readily seen. Thus with two terms, x and y, we have four combinations; $xy, x\bar{y}, \bar{x}y, \bar{x}\bar{y}$. Introduce the term z and we at once split up each of these four into z and not-z parts, and so double their number. Provided our diagrams are so contrived as to indicate this, they will precisely correspond, in every relevant respect, to the table of combinations of letters. (Venn 1881, p. 113–14)

Now, in order to show how Venn employs diagrams of overlapping compartments to illustrate the Boole's procedure of development through a visualisation of the relations between classes, the syllogisms c**AmEnE**s and d**ImArI**s, which belong respectively to the mood **A E E** and **I A I** in the fourth figure, should be worked out. But, before proceeding to employ the diagrammatic method, it should be noted that for Venn, in the diagrammatic representation of proposition, the symbols stand for compartments. Then, when it is required

---

[2] For a clear exposition of Euler's endeavour as well as an analysis of its defects Venn's *Symbolic Logic* may be consulted. A brief historic of some previous attempts to design the geometric illustration of proposition before and after Euler is also given in the last chapter of this treatise. On these diagrammatic representation of logical process, see also Martin Gardner: *Logic Machines and Diagrams* (1958). For a criticism of Euler's circles, see L. S. Stebbing, *A Modern Introduction to Logic*, (1950, pp. 72–78).

to determine what combinations are emptied out by any given proposition, they are just shaded out, so as to indicate that these compartments are unoccupied. When it is required to indicate that a compartment is occupied, a star ★ is placed inside it. If it is not sure whether a star ★ belongs in one compartment or an adjacent one, then it is put on the line between the two compartments.

I shall proceed to carry out the diagrammatic scheme. Let us draw first three circles, which intersect as follows:

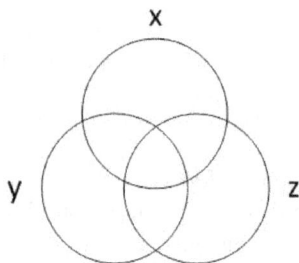

The circles are marked x, y, and z. Each one is cut up into four parts, and each common part of two circles into two parts. There are eight compartments. All the points inside circle x are members of the compartment x. All point outside the same circle are not-x. And likewise with the other circles. The circles overlap in such way that, if each compartment is marked with appropriate symbols to indicate its members, then a compartment is obtained for every possible combination of the three symbols and their negation. Hence, there are eight compartments, which correspond to the eight combinations given by the three symbols:

$$xyz,\ xy\bar{z},\ x\bar{y}z,\ x\bar{y}\bar{z},\ \bar{x}yz,\ \bar{x}y\bar{z},\ \bar{x}\bar{y}z,\ \bar{x}\bar{y}\bar{z}.$$

Each compartment has a symbolic name ready provided for it, and with a finger the compartment referred to can be indicated. The schemes of symbols and compartments agree in their elements being mutually exclusive and collectively exhaustive. The method may be extended to any number of terms. A pair of terms give 2^2 combinations, three 2^3 combinations, four 2^4 combinations or 16, and so on.

Some Further Developments of Boole's Logical Calculus 115

The diagram representing three terms is drawn as follows:

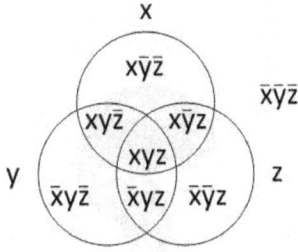

The compartment outside of all three circles represents the compartment of $\overline{xyz}$, or all those things which are not members of any of the three classes.

Accordingly, the first syllogistic inference, c**Am**E**n**E**s**, may be represented by means of the diagrammatic method,

All x is y
No y is z
$\therefore$ No x is z.

The first premise, 'All x is y', means that the compartment of things which are x and not-y is unoccupied. All compartments in which these two terms are found must then be shaded out thus:

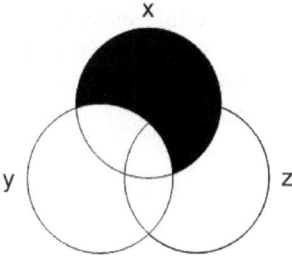

The second premise is 'No y is z', which tells us that all compartments containing the combination yz are unoccupied; the diagram is further shaded out like this:

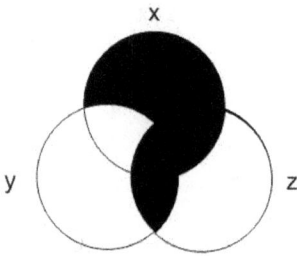

At this level, it must be seen that a valid conclusion concerning the relation between x and z can be drawn. For all compartments containing both x and z are unoccupied; therefore 'No x is z' is inferred. If it is supposed that x is not an unoccupied compartment, then 'Some x is not z' can also be inferred.

Regarding the second syllogistic inference, d**Im**A**r**I**s**, it shows how particular premises can be represented as

$$\text{Some } x \text{ is } y$$
$$\text{All } y \text{ is } z$$
$$\therefore \text{Some } x \text{ is } z$$

The first premise of this syllogistic inference, 'Some x is y', requires a star ★ on the edge of the z circle, for it is not known which of the two compartments (or perhaps both) may be occupied. The following diagram shows the pictorial representation of the first premise:

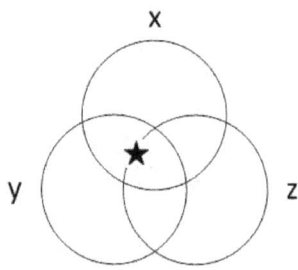

The second premise, 'All y is z', empties out one of these compartments. So we must put the star ★ in the occupied compartment as illustrated in this diagram:

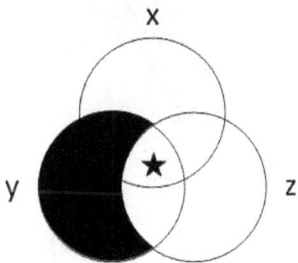

A relation of x to z is now readily detected. From the two premises, the conclusion, 'Some x is z', must be drawn.

Thus, Venn's diagrammatic method allows the visualisation of the logical operations of Boole's calculus of classes. As a result, even a philosopher without any mathematical background can 'see' now what Boole's logical calculus is all about. The method is somehow similar to the decision procedure so far explained (see subsection **3.4.2**). It enables one to test the validity of syllogisms.

The diagrammatic method can be employed so as to represent a disjunctive relation. For instance, the proposition, all x is either y or z in which 'or' is interpreted in the inclusive sense of 'either or both,' yields this picture:

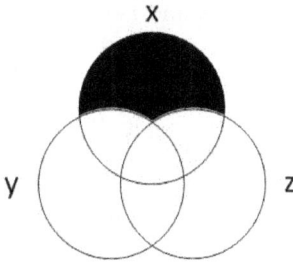

For the 'exclusive' disjunction the following diagram is obtained:

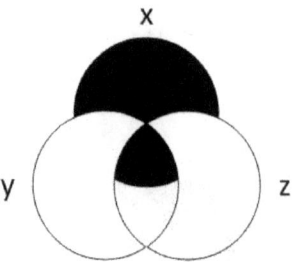

The method can be extended so as to carry out four terms. Venn notes that with four terms the most simple and symmetrical diagram is drawn by making four ellipses intersect one another in the following manner,

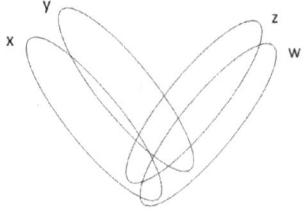

He then shows the scope and power of the diagrammatic method whilst working out the following premises,

Every y is either x and not z, or z and not x.

Every wy is either both x and z or neither of the two.

All xy is either w or z, and all yz is either x or w.

It is required to work out any obvious conclusion, which follows from them. Venn finds that these propositions involve the elimination of the following classes: yxz, $y\overline{xz}$ by the first premise; $wyx\overline{z}$, $wy\overline{x}z$ by the second premise; and $xy\overline{wz}$, $yz\overline{xz}$ by the third premise.

Some Further Developments of Boole's Logical Calculus

When the corresponding compartments in the diagram are shaded out, this result is obtained,

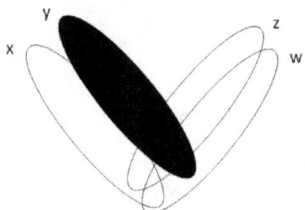

Subsequently, Venn says,

> It is then clear at a glance that the collective effect of the given premises is just to deny that there can be any such class of things as y in existence, though they leave every one of the remaining eight combinations perfectly admissible. This, then, is the diagrammatic answer to the proposed question. (Venn 1881, p. 130)

Although Venn does not elaborate specifically his method, so as to deal with propositional calculus, he recognises that the diagrams can be employed for working out problems in this domain. In Chapter XVIII of *Symbolic Logic*, entitled 'Class Symbols As Denoting Propositions', he sets out to interpret the use of his symbols in a way in which they 'stand for propositions, i.e. not for the propositions themselves, but for their truth or falsehood' (Venn 1881, p. 43). In the light of Boole's calculus of secondary propositions, he carries out all the required material that allows an interpretation of the class calculus as propositional calculus. There follows the possibility of diagramming these propositions when their premises are not made more complex by parenthetical assertions. But in order to employ the Venn circles for propositional logic we must give a different interpretation. Thus, each circle represents a proposition which may be true or false rather than a class which may or may not have members. The marks on the compartments indicate possible or impossible combinations of true and false values of the terms. A compartment is shaded out to show that it is an impossible combination of truth-values. A compartment left blank indicates a permissible combination. The combination $\overline{XYZ}$ is represented as a small circle outside the other three for simplifying the shading of this compartment when required.

Here is the diagrammatic scheme of the propositional calculus:

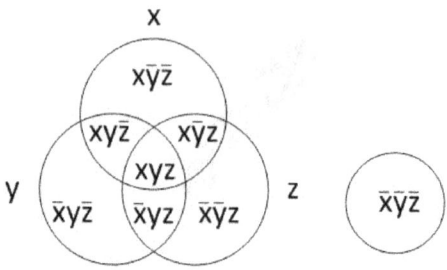

The method can be further illustrated by working out the conditional proposition, 'If X then Y' or $X \Rightarrow Y$. The truth-table of the conditional shows that there are four possible true and false combinations, (TT, TF, FT, FF), amongst which only the combination TF is not valid.

Likewise the Venn diagram for the conditional represents the four possible combinations of truth-values (XY, $X\overline{Y}$, $\overline{X}Y$, $\overline{XY}$), amongst which only the compartment $X\overline{Y}$ indicates an impossible combination of truth-values. Hence, in the above diagram, all compartments containing $X\overline{Y}$ are shaded out. As a result, a diagram equivalent to the diagram for the universal affirmative proposition 'All x is y' is obtained. It is represented as follows:

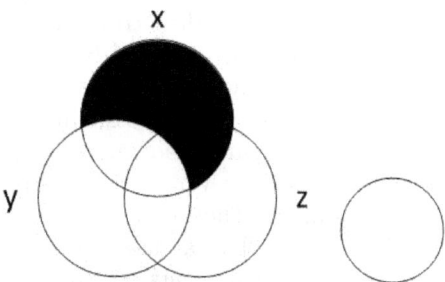

It should be mentioned that, at the time when Venn devised the diagrammatic method, other rectangular diagrams were drawn by Allen Marquand, Alexander Macfarlane, and Lewis Carroll[3]. All these diagrams can be used for

---

[3] Allen Marquand released his paper titled "A Logical Diagram for n Terms" which appeared in the *Philosophical Magazine*, vol. 12, October, 1881. Macfarlane's "The Logical Spectrum" is published in *Philosophical Magazine*, vol. 19, 1885. Lewis Carroll's diagram appeared in his book, *The Game of Logic*, 1886.

Some Further Developments of Boole's Logical Calculus

the calculus of classes and propositions because of the isomorphism between the two calculus.

Venn's diagrams have been employed to illustrate the concepts, relations and procedures of Boole's logical calculus. But, although the diagrams are practical for many purposes, they are only a means of illustration which should not spare acquaintance with the algebra. Moreover, an important weakness of the method is that the diagram is appropriate only with an upper limit of four classes. Hence, if Boolean algebra is restricted to its application in Venn's diagrams, then it becomes incomplete. Furthermore, the mathematical imagination may be dampened when it relies too much upon diagrams. However, a full understanding of the algebra, as a system of pure logic, is best achieved by means of a postulational method, which carries out a logical or mathematical system.

4.2 Modern Algebra of Logic

Modern algebra of logic arises from the work of Boole's followers, who have straightened out his logic of classes by jettisoning what was not completely intelligible in the system. As a result, they study an algebraic structure, that is a set with one or more operations defined on it, of which the natural numbers with addition and multiplication constitute an example. Such an abstract structure is what is called 'Boolean Algebra.' As Tarski defines it,

> Boolean Algebra, also called the *algebra of logic*, is a formal system with a series of important interpretations in various fundamental departments of logic and mathematics. The most important and best known interpretation is the calculus of classes. (Tarski 1956, p. 320)

Thus, there are several equivalent systems of primitive symbols and axioms, which concur in the setting up of the abstract calculus. But Boolean algebra has been interpreted, for the most part, as applying to classes. Then, as an example of these systems, let us construe Boolean algebra as a structure including a set B, two binary functions ∩ (conjunction or intersection) and ∪ (disjunction or union) on B, plus one unary function (complementation) on B, and two distinct elements 0 (the null-element) and 1 (the unit element) of B, satisfying the following axioms, for all $x, y, z \in B$

(i) $x \cap y = y \cap x$ and $x \cup y = y \cup x$
(ii) $(x \cap y) \cap z = x \cap (y \cap z)$ and $(x \cup y) \cup z = x \cup (y \cup z)$
(iii) $1 \cap x = x$ and $0 \cup x = x$
(iv) $x \cap \bar{x} = 0$ and $x \cup \bar{x} = 1$
(v) $x \cap (y \cup z) = (x \cap y) \cup (x \cap z)$ and $x \cup (y \cap z) = (x \cup y) \cap (x \cup z)$
(vi) $x \cap x = x$

In order to avoid confusing Boolean algebra with the arithmetical operation of addition the symbol '∪' is here employed as the inclusive disjunction or union that corresponds to Boolean 'addition' sign '+' when it is interpreted inclusively, and the symbol ∩ as the conjunction or intersection that corresponds to Boolean multiplication sign. This notation is taken from the calculus of classes of Whitehead and Russell in *Principia Mathematica*. The notation of the negation belongs to Sheffer and Huntington, who put the bar over the class symbol except where a parenthesis is negated. Now, (i) is called the *laws of commutativity*, (ii) the *laws of associativity*, (iii) & (vi) the *laws of tautology*, (iv) the *laws of complementation* and (v) the *laws of distributivity*.

There may be added some other fundamental laws such as the *law of unity*:

$$x \cup x = x$$

and the *laws of absorption*:

$$x \cup (x \cap y) = x$$
$$x \cap (x \cup y) = x,$$

which were formulated by Jevons.

Since a conjunction can be expressed as a disjunction if the negation of the classes is performed, the following equivalences called *de Morgan's theorems*[4] are fundamental laws of Boolean algebra:

$$(x \cup y) = -(\bar{x} \cap \bar{y}),$$
$$(x \cap y) = -(\bar{x} \cup \bar{y}).$$

The ordinary assumption that a double negation is equivalent to an affirmation is also a fundamental law of the algebra. It is called the *law of double negation*:

$$\bar{\bar{x}} = x.$$

---

[4] By referring to the Venn diagram, as an application of Boolean algebra, these theorems can be readily shown to hold. But this does not prove them to be true. However, they are a special case of the law of duality which can be proved. See C. I. Lewis, *A Survey of Symbolic Logic* (1960), pp. 123–126; W. V. Quine, *Methods of Logic* (1974), pp. 59–63; and A. Church, *Introduction to Mathematical Logic* (1956), pp. 106–108, 201–205.

Some Further Developments of Boole's Logical Calculus

Finally, the *law of expansion* may be expressed. It allows a class symbol to be developed by reference to another class when conjoining it first with the other class, and then with the negation of the other class, and then disjoining the two conjunctions. Hence we have,

$$x = (x \cap y) \cup (x \cap \bar{y}).$$

All these fundamental laws allow one to perform Boolean operations by working out problems in the algebra. In this direction, 0 stands for the empty class, and when a class is empty, it is equated to 0. On the other hand, when a class has members, it is not equal to 0. It follows that the four propositions of traditional logic are expressed in Boolean methods by equations or inequations (see also subsection **4.1.1**). Thus, we obtain the following scheme:

A.	All x is y	$x \cap \bar{y} = 0$
E.	No x is y	$x \cap y = 0$
I.	Some x is y	$x \cap y \neq 0$
O.	Some x is not y	$x \cap \bar{y} \neq 0$

In order to handle the equations and the inequations, there are four rules of inferences in Boolean algebra. The first two rules concern the equations and are stated as follows:

(i) If $x = 0$ and $y = 0$, then $x \cup y = 0$.
(ii) If $x \cup y = 0$, then $x = 0$ and $y = 0$.

The two other rules concern the inequations:

(iii) If $x \cap y \neq 0$, then $x \neq 0$ and $y \neq 0$.
(iv) If $x \cup y \neq 0$ and $x = 0$, then $y \neq 0$.

Rule (i) states that if two classes are empty, then their disjunction is empty. Rule (ii) is the inverse of (i), that is, if the disjunction of two classes are empty, then each of the disjoined classes is empty. Rule (iii) states that if the conjunction of two classes has at least one member, then each of the conjoined classes has at least one member. Rule (iv) states that if the disjunction of two classes has at least one member, and if one of the classes is empty, then the member belongs to the other class. It may be worth noting that rule (iv) is not the inverse of rule (iii). The inverse of rule (iii) does not hold. From $x \neq 0$ and $y \neq 0$, it cannot be inferred that the members in either case is in the conjunction of x and y. Hence $x \cap y \neq 0$ cannot be asserted on the basis of $x \neq 0$ and $y \neq 0$.

Let us now illustrate how the algebraic methods perform in the calculus of classes by working out the syllogism c**El**Ar**Ent**, which belongs to the mood **E A E** in the first figure. The premises are expressed as follows,

$$y \cap x = 0$$
$$z \cap \bar{y} = 0.$$

It is required to ascertain the relation between x and z. Thus, according to the *law of expansion*, which states that a class symbol is expanded by reference to another class by conjoining it first with the other class, and with the negation of the other class, and then disjoining the two conjunctions, we develop the first premise with respect to z. Hence we obtain:

$$((y \cap x) \cap z) \cup ((y \cap x) \cap \bar{z}) = 0.$$

In like manner we develop the second premise with respect to x, and obtain:

$$((z \cap \bar{y}) \cap x) \cup ((z \cap \bar{y}) \cap \bar{x}) = 0.$$

According to the above first rule of inference (i), we may add the premises, and obtain:

$$(((y \cap x) \cap z) \cup ((y \cap x) \cap \bar{z})) \cup (((z \cap \bar{y}) \cap x) \cup ((z \cap \bar{y}) \cap \bar{x})) = 0.$$

The *law of associativity* allows us to rewrite the first and the third constituents of this long disjunction as follows:

$$(z \cap x) \cap y \text{ and } (z \cap x) \cap \bar{y}.$$

By the second above rule of inference (ii) applied twice and the first rule of inference, we can expunge the second and the last constituents, and thus we obtain the equation:

$$((z \cap x) \cap y) \cup ((z \cap x) \cap \bar{y}) = 0.$$

It is now easy to eliminate y by applying the Boolean rule of elimination, that is, if an expression is put into the form $(x \cap y) \cup (z \cap \bar{y}) = 0$, then the result of elimination will be $x \cap z = 0$. Hence from our formula we obtain:

$$z \cap x = 0,$$

which is the valid conclusion of the syllogism c**El**Ar**Ent**.

Some Further Developments of Boole's Logical Calculus

As well, we could get $z \cap x = 0$ straight from the law of expansion

$$x = (x \cap y) \cup (x \cap \bar{y}),$$

with $z \cap x$ substituted for x.

We may carry out all syllogistic inference containing universal premises in a similar way. The procedure consists of developing each premise so as to involve all classes, and then adding the premises. In the case of a valid form of syllogism, we will notice that there are two constituents in the disjunction such that the middle class may be eliminated from them. But if the form of the syllogism is invalid, then we will not have two constituents such that the middle term may be eliminated, and thus no conclusion may be drawn.

However, regarding a particular premise, this procedure changes, for the particular premise is expressed by an inequation, and thus the premises cannot be added. As an illustration, let us consider the syllogism dImArIs, which belongs to the mood **I A I** in the fourth figure. The premises may be symbolised as follows:

$$x \cap y \neq 0$$
$$y \cap \bar{z} = 0.$$

In order to eliminate y and determine the relation between x and z, we develop the first premise with respect to z:

$$((x \cap y) \cap z) \cup ((x \cap y) \cap \bar{z}) \neq 0.$$

Next, we develop the second premise with respect to x:

$$((y \cap \bar{z}) \cap x) \cup ((y \cap \bar{z}) \cap \bar{x}) = 0.$$

According to the second rule of inference (ii), we may infer from the above disjunction that

$$(y \cap \bar{z}) \cap x = 0.$$

By the law of associativity we may rewrite this conjunction as,

$$(y \cap \bar{z}) \cap x = (x \cap y) \cap \bar{z}.$$

As may be noticed, it turns out that $(x \cap y) \cap \bar{z}$ is one of the constituents in the expansion of the first premise that is,

$$((x \cap y) \cap z) \cup ((x \cap y) \cap \bar{z}) \neq 0.$$

This disjunction is not empty, but $(x \cap y) \cap \bar{z}$ is empty according to the above procedure, which proceeds by the development of the second premise with

respect to x, the application of the second rule of inference (ii) and the use of the law of associativity. Hence, the fourth rule of inference (iv) allows us to infer that
$$(x \cap y) \cap z \neq 0.$$
Now, according to the third rule of inference (iii), if $(x \cap y) \cap z \neq 0$, then
$$x \cap z \neq 0.$$
This is the valid conclusion of the syllogism dImArIs.

The same procedure may be employed to work out all valid syllogisms containing a particular premise. We first proceed by developing the two premises where the universal premise shows that one of the conjunctions in the particular premise is empty. Then by the fourth rule of inference, we may infer that the other conjunction in the particular premise is not empty. This will leads to a valid conclusion by the third rule of inference.

I shall spare the reader an illustration containing more than three classes, though the algebraic methods work out any number of classes involved in a logical problem.

It should be pointed out that, in working out the above two syllogisms, we came across formulas, such as
$$(((y \cap x) \cap z) \cup ((y \cap x) \cap \overline{z})) \cup (((z \cap \overline{y}) \cap x) \cup ((z \cap \overline{y}) \cap \overline{x})) = 0,$$
which is called *disjunctive normal form*. For, in this disjunction, the negation sign applies only to individual variables, which are conjoined, and then the conjunctive sets are disjoined. Boole's *law of development* (expansion into constituents) exemplifies a formula in disjunctive normal form (see subsection **3.2.1**).

For instance, in Boole's notation the formula is written as
$$yxz + yx(1-z) + z(1-y)x + z(1-y)(1-x) = 0.$$

Now, when Boole's exclusive disjunction is substituted for the inclusive symbol $\cup$ and the multiplication of classes represented by the symbol $\cap$, the formula corresponds to the above disjunctive normal form.

We may put a formula into disjunctive normal form by using successively *de Morgan's theorems*, the *laws of distributivity* and the *law of double negation*. Whilst dealing with the syllogistic inference, we have employed such a disjunctive normal form, which comes out with the expansion of the premises of the given syllogism to obtain all classes involved. Thus, the rules of inference and the laws of the system have allowed us to draw the valid conclusion. As a decision procedure, the disjunctive normal form enables us to work out the decision problem for the calculus of classes.

In the interpretation of Boolean algebra as applying to classes,[5] the symbol '1' stands for the universal class, and the expansion of the disjunctive normal form for classes yields the universal class. The symbol '0' stands for the null class, and the denial of the Boolean expansion of the disjunctive form for classes is the null class. The symbol ∪ allows an operation on two classes which yields a new class containing as its members those, and only those, things which belong to at least one of the two classes. The operation is called addition of classes, and the resulting class is the sum or union of the two classes. The symbol ∩ is an operation on two classes called multiplication of classes. It consists in forming a new class whose members are those, and only those, things which belong to both classes. The resulting class is the product or intersection of the two classes.

Boolean algebra applies as well to several mathematical domains, e.g., a geometrical application to overlapping plane compartments of space and set theory.In Huntington's eyes, the most familiar example of a Boolean algebra is the following:

$K =$ the class of regions in a square (including the null region and the whole square);

$a + b =$ the smallest region which includes both a and b;

$a' =$ the region complementary to a with respect to the square;

$ab =$ the region common to a and b.

Here the relation '$a < b$ means a is included in b' (Huntington 1933, p. 278). Venn's diagrams are an illustration of such an application of Boolean algebra.

It is worth noting that Boolean algebra is set up as a logical structure, or an abstract system. Accordingly, whether or not an interpretation is given is out of play in the structure of the system. The various interpretations which may be given do not define the logical system.

---

[5] As might be expected, Boolean algebra can also be interpreted as applying to propositional forms defined in terms of their truth values. For a thorough development of this application, I shall refer the reader to chapters three and four of Lewis's book: *A Survey of Symbolic Logic*, and chapters fourteen and fifteen of Newton Lee's *Symbolic Logic*, which both constitute the main source of inspiration when carrying out the above description of Boolean algebra.

4.3 Axiomatic Methods for 'Boolean Algebras'

The most important development of Boole's logical calculus is the setting up of axiomatic methods for the calculus. Although Boole did not lay out a system of axioms, it is undeniable that his algebra of logic provides the basis for a method in which the principles are cast into the form of postulate sets. Indeed, in the beginning of the twentieth century, Huntington and Sheffer presented explicitly the implicit structures embedded in Boole's computational procedure, and thus opened a new avenue, namely the metamathematical inquiry into the algebra of logic. In what follows, I shall give an account of the axiomatic methods in their respective papers on "A Set of Five Independent Postulates for Boolean Algebras"and "Sets of Independent Postulates for the Algebra of Logic".[6]

4.3.1 Huntington: Sets of Independent Postulates for Boolean Algebra

In 1904 and 1933, Huntington published two papers, in which he developed six sets. I shall be limiting myself to giving an expository account of the first and the fourth sets. Huntington introduces the paper of 1904 by stressing that

> The algebra of symbolic logic, (...) although originally studied merely as a means of handling certain problems in the logic of classes and the logic of propositions, has recently assumed some importance as an independent calculus; it may therefore be not without interest to consider it from a purely mathematical or abstract point of view, and to show how the whole algebra, in its abstract form, may be developed from a selected set a fundamental propositions, or postulates, which shall be independent of each other, and from which all the other propositions of the algebra can be deduced by purely formal processes. (Huntington 1904, p. 288)

---

[6]These papers are published in *Transactions of the American Mathematical Society* vol. 14 (1913) for Sheffer; vol. 5 (1904), pp. 288–309, and vol. 35 (1933), pp. 274–304, for Huntington. Others axiomatic methods have been also constructed for Boolean algebra. A brief bibliography of these sets of postulates is given by Huntington in his paper "New Sets of Independent Postulates for The Algebra of Logic" published in *Transactions of the American Mathematical Society* vol. 35 (1933) p. 27. Later, Tarski published a paper (1935) containing a study of some alternative systems of Boolean algebra. As a result, he brought out several equivalent sets of postulates for an extended (or complete) system of Boolean algebra, which deals with two operations, namely the logical sum and the logical product applicable to all elements of a certain set of elements, or universe of discourse, denoted by 'B'. The paper can be found in his book *Logic, Semantics, Metamathematics* (1956, pp. 320–334).

Some Further Developments of Boole's Logical Calculus 129

Thus, he sets out to build up an axiomatic system for the algebra of logic regardless to its possible interpretations.

In Huntington's eyes, a postulate set is regarded as a system, in which the primitive ideas constitute the set K and may be called 'a class (K) of elements' $(a, b, c, \ldots)$.' The primitive relations are the two 'rules of combination', $+$ and $\times$ (Huntington 1904, pp. 288–289). The postulates bring the K-elements and the rules of combination together into a system.

Then he states the object of his paper as follows:

> Having chosen a set of fundamental concepts and a set a fundamental proposition for each of the three sections, I show, first, that the fundamental propositions of each set are consistent (and independent); and secondly, that the fundamental concepts of each section can be defined in terms of the fundamental concepts of each of the other sections, while the fundamental propositions of each section can be deduced from the fundamental propositions of each of the other section. Then we may say, first, that each section determines a definite algebra, and secondly, that the three algebras are equivalent. (Huntington 1904, pp. 290–291)

The first section carries out Huntington's first set, which is the closest to Boole's system in that it considers the two operations of addition and multiplication as primitive relations. But, Huntington distinguishes the inclusive disjunction from the arithmetical operation by placing the sign in a circle, $\oplus$, and the conjunction by using a dot in a circle, $\odot$. He also employs the symbols $\wedge$ and $\vee$, which resemble an empty glass and a full glass, so as to facilitate the respective interpretation of them as 'nothing' and 'everything'. He borrows them from Peano's *Formulaire de Mathématiques*. In the paper of 1933, he changed the symbolism whilst reproducing the first set. Thus, the original $\wedge$, $\vee$, and $\bar{a}$ are replaced respectively by z, u, and a'; and the circle around $+$ and $\times$ are omitted. For convenience I shall use the latter symbolism in the following tabulation of the postulates of the first set. The postulates define a class K with elements, a, b, etc.

Ia. $a + b$ is in K whenever a and b are in K.
Ib. $a \times b$ is in K whenever a and b are in K.
IIa. There is an element z such that $a + z = a$ for every element a in K.
IIb. There is an element u such that $a \times u = a$ for every element a in K.
IIIa. $a + b = b + a$ whenever a, b, $a + b$, $b + a$ are in K.
IIIb. $a \times b = b \times a$ whenever a, b, $a \times b$, and $b \times a$ are in K.
IVa. $a + (b \times c) = (a + b)(a + c)$ whenever a, b, c, $a + b$, $a + c$, $b \times c$, $a + (b \times c)$, and $(a + b)(a + c)$ are in K.
IVb. $a \times (b+c) = a \times b + a \times c$ whenever a, b, c, $a \times b$, $a \times c$, $b+c$, $a \times (b+c)$, and $(a \times b) + (a \times c)$ are in K.
V. If the elements z and u in postulates **IIa** and **IIb** exist and are unique, then for every element a in K there is an element a' such that $a + a' = u$ and $a \times a' = z$.
VI. There are at least two elements, x and y, in K such that $x \neq y$.

It should be noted that Huntington acknowledges that in the postulates **Ia–V**, he follows closely Whitehead, who gave a complete general theory of algebra and presented what he called 'the formal laws' of the algebra in his *A Treatise on Universal Algebra*, (1898, vol. I pp. 35–37). In it, he set out 'to exhibit the new algebras, in their detail, as being useful engines for the deduction of propositions; and in their several subordination to dominant ideas, as being representative symbolisms of fundamental conceptions' (Whitehead 1898, p. viii).

In the above postulates, there are two particular K-elements, the empty class and the universal class, and one particular primitive operation, the negation. The system may be described as involving the K-elements a, b, c,..., which can be interpreted either as classes or propositions; and the particular K-elements z, the empty class, and u, the universal class or falsity and truth. Huntington formulates a rule of inference based upon the symbol '='. The particular primitive operation, the negation, is considered as undefined.

The undefined rules of combination are addition (disjunction), multiplication (conjunction) and negation. There are ten postulates, which show how the K-elements a, b, c,..., z and u, and the rules of combination $+$, $\times$ and $-$ can be put together into meaningful propositions of the system.

The Postulates **IIIa** and **IIIb** are well known as the law of commutativity for addition and multiplication.

As well, the postulates **IVa** and **IVb**, which are the laws of distributivity for addition and multiplication.

Postulates **IIa**, **IIb**, and **V** postulate the existence of the empty class, the universal class and the negation.

Some Further Developments of Boole's Logical Calculus 131

Postulates **Ia** and **Ib** state that the addition of any two K-elements is itself a K-element, and the multiplication of any two K-elements is itself a K-element. They are what Sheffer calls K-closing postulates.

Postulate **VI**, which is the last postulate of the set, is intended to ensure that the system is not empty by stating that there are at least two distinct K-elements. The postulate is included only to keep the system from being vacuous.

I shall spare the reader the metamathematical apparatus, upon which Huntington relies in order to prove the consistency and the independence of the first set. However, let us point out that he demonstrates the consistency of the postulates by exhibiting a system $(K, +, \times)$ in which K, $+$, and $\times$ are so interpreted that all the postulates are satisfied. 'For then the postulates themselves, and all their consequences, will be simply expressions of the properties of this system, and therefore cannot involve contradiction (since no system which really exists can have contradictory properties)' (Huntington 1904, p. 293). As an illustration, he gives the system: $K =$ the class of regions in the plane including the 'null region' and the whole plane; $a + b =$ the 'logical sum' of a and b; $ab =$ the 'logical product' of a and b.

The ten postulates are also independent, inasmuch as none of them can be drawn from the other nine. This can be demonstrated by exhibiting for each postulate, a system $(K, +, \times)$, which satisfies all the other postulates, but not the one in question. 'This postulate, then, cannot be a consequence of the others; for if it were, every system which had the other properties would have this property also, which is not the case' (Huntington 1904, p. 296).

Huntington's fourth set is presented in the paper of 1933 in which he builds up three more sets. The paper is suggested by the developments made in the setting up of postulates for Boolean algebra. Amongst these developments, he notes the set of postulates given by Sheffer in 1913, which are expressed in terms of $(K, /)$ where the 'stroke' represents another binary operation called 'rejection,' such that $a/b = (a + b)'$. Then he points out the two sets given by Bernstein in 1914, which are expressed in terms of $(K, -)$ where the '$-$' represents another binary operation called 'exception,' such that $a - b = a \times b'$; and in terms of $(K, \div)$, where the '$\div$' indicates a binary operation called 'adjunction,' such that $a \div b = a + b'$.

Huntington was particularly struck by the expression of the primitive propositions of Section A of the *Principia Mathematica* (1910) in terms of a class called the class of 'elementary propositions,' a binary operation called 'disjunction,' and a unary operation called 'negation'. In 1931, this stimulated Bernstein to show how these primitive propositions can be expressed in abstract mathematical form in terms of $(K, +, ')$. Subsequently, there occurred a discussion about the relation between the theory of the *Principia* and the theory of Boolean algebra.[7] According to Huntington, there follows that 'it becomes a matter of interest to construct a set of independent postulates for Boolean algebra explicitly in terms of $(K, +, ')$, for comparison with the *Principia*' (Huntington 1933, p. 276). I shall be saying a little more about this comparison when commenting upon the following fourth set of Huntington:

4.1. If a and b are in the class K, then $a + b$ is in the class K.
4.2. If a is in the class K, then a' is in the class K.
4.3. $a + b = b + a$.
4.4. $(a + b) + c = a + (b + c)$.
4.5. $a + a = a$.
4.6. $(a' + b')' + (a' + b)' = a$.
4.7. $ab := (a' + b')'$.
4.8. $ab + ab' = a$.

This set of postulates is expressed in terms of $(K, +, ')$, K is a class of elements $a, b, c, \ldots$; $a + b$ denotes the result of a binary operation called *logical addition;* and a' denotes the result of a unary operation called *logical negation.* The set shows how the whole system can be built up with the operation of addition, or disjunction, and the negation as the sole primitive relations, or rules of combination. It should be noticed that this time Huntington does not involve the empty class and the universal class in the postulates. He slips into the system the universal class, called the 'universe' element of the system, and the empty class, called the 'zero' element of the system, through explicit definitions, and then gives the proof of the theorems about them. The set contains eight postulates, but he later deleted **4.5**, for it can be deduced from the others (*Transactions of the American Mathematical Society*, vol. 35, p. 557). By aid of the definition **4.7** the postulate **4.6** can be replaced by **4.8** Huntington recognises that the set four 'appears to be the simplest and most "natural" of all the sets of postulates for Boolean algebra' (Huntington 1933, p. 276).

Moreover, in Appendix I of the paper, Huntington seeks the connection between Boolean algebra and the *Principia* through a comparison between set

---

[7] In his paper of 1913, Sheffer applied his set to the primitive logical constants given by Whitehead and Russell in the *Principia*, and had made it possible 'to reduce, by one each, the number of primitive ideas and of primitive propositions used in the *Principia* for the foundation of logic.' (pp. 486–488) In Appendix II of his paper of 1933, Huntington himself carried out a revision of the primitive propositions of the *Principia* (Section A), without using the equality sign, or the Boolean notation '= 1,' and constructed a set of independent postulates for *Principia Mathematica* (pp. 301–304).

Some Further Developments of Boole's Logical Calculus

four and the elementary propositions set forth in Section A of the *Principia*. So as to make the comparison, he first quotes the propositions in question from the second edition of the *Principia*.

1.71. If p and q are elementary propositions, then $p \lor q$ is an elementary proposition.

1.7. If p is an elementary proposition, then $\sim p$ is an elementary proposition.

4.31. $\vdash : p \lor q . \equiv . q \lor p$
4.33. $\vdash : (p \lor q) \lor r . \equiv . p \lor (q \lor r)$
4.25. $\vdash : p . \equiv . p \lor p$
4.5. $\vdash : p.q . \equiv . \sim(\sim p \lor \sim q)$
4.42. $\vdash : .p. \equiv : p.q. \lor .p. \sim q$

Then, Huntington notes that if we call the class of 'elementary propositions' the class K, and write $p + q$ for $p \lor q$, and p' for $\sim p$, then 'these propositions are precisely the same as the postulates of our fourth set of Boolean algebra, except that the sign $\equiv$ occurs in place of the sign =' (Huntington 1933, p. 299). Consequently, he goes on to examine the properties of the sign $\equiv$ as used in the *Principia*, in comparison with the postulates he lays down governing the use of the equality sign = in his fourth set. This leads him to distinguish between what he calls the formal statements and the informal statements in the *Principia*. Amongst the informal statements he cites **4.22**, which states that '$\equiv$' denotes 'the relation of equivalence,' and that the relation of equivalence is reflexive **4.2**, symmetrical **4.21** and transitive **4.22**'. This comes to the same thing as the three following postulates of the set for =, which he gives to 'express Boolean algebra in terms of the four undefined concepts $(K, +, ', =)$' (Huntington 1933, p. 280).

A. If a is in the class K, then $a = a$.
B. If $a = b$, then $b = a$.
C. If $a = b$ and $b = c$, then $a = c$.

Therefore, if the above mentioned informal statement is considered as a valid part of the theory of the *Principia*, Huntington can state the following theorem:

> With respect to $(K, \lor, \sim, \equiv)$, the informal system of the *Principia* is a Boolean algebra. (Huntington 1933, p. 300)

One implication of this theorem is that Boole was certainly a pioneer in the progress of modern formal logic since his algebraical procedures survive in modern abstract algebra, and particularly in Boolean algebra. As has been admitted in the *Principia,* Russell and Whitehead were 'concerned with rules analogous, more or less, to those of ordinary algebra. It is from these rules that the usual "calculus of formal logic"starts' (Whitehead and Russell 1927, vol. I, p. 115). Accordingly, Boole's calculus of classes is, indeed, the historical point of departure of the *Principia,* which encompasses the general theory of classes that develops the whole of arithmetic, including algebra, analysis, and so on, as a part of pure logic.

In addition, it is interesting to note that for Quine, the Boolean algebra of unions, intersections, and complements, which he construes as a 'virtual theory of classes', is logic, pure logic in disguise, whereas Frege, Whitehead, and Russell's general theory of classes, which admit '0' as a genuine predicate, and classes as values of quantifiable variables, is embarked on a substantive mathematical theory (Quine 1970, p. 72).[8] Quine even argues that what is done in the name of set theory is quite within the reach of Boolean notation of unions, intersections and complements, which 'merely does in another notation what can be done in that part of logic of quantification which uses only one-place predicate letters. The variables in Boolean algebra are unquantified and can be read as schematic one-place predicate letters' (Quine 1970, p. 69).

4.3.2 Sheffer: A Set of Five Independent Postulates for Boolean Algebras

Sheffer is the first to suggest the name of 'Boolean algebras' in the paper, "A Set of Five Independent Postulates for Boolean Algebras", written in 1913. He produces the most economical set of postulates in that it deals only with one primitive operation:

1. There are at least two distinct K-elements.

2. Whenever a and b are K-elements, a/b is a K-element and a/a is denoted by a'.

3. Whenever a and the indicated combinations of a are K-elements, $(a')' = a$.

4. Whenever a, b, and the indicated combinations of a and b are K-elements, $a/(b/b') = a'$.

---

[8]It is not my purpose here to elaborate this point related to the scope of logic and to the reductionist programme of Frege and his followers (see subsection **6.3.2** and Quine 1970, pp. 61–79).

Some Further Developments of Boole's Logical Calculus 135

5. Whenever a, b, c, are the indicated combinations of a, b, and c are K-elements, $(a/(b/c))' = (b'/a)/(c'/a)$.

The set assumes a class K—elements $a, b, c, \ldots$, a binary K-rule of combination the stroke, '/', and five postulates. It differs from the previous sets in that it contains a smaller number of postulates, and does not involve existence postulates for the empty class, the universal class, or negation. He uses negation in some of the postulates, but he defines it as $a' := a/a$.

Later, when he applies the set to primitive logical constants given in the *Principia*, he defines disjunction in term of *rejection*[9] as follows:

Defⁿ: For any two elementary propositions, p and q,

$$p \cup q = (p/q)/(p/q).$$

He draws the empty class and the universal class from the postulates, even if they do not figure in the statement of his postulates. The theorems of the empty class and the universal class are:

IIa. There is a K-element z such that for any K-element a, $(a/z)' = a$.

IIb. There is a K-element u such that for any K-element a, $a'/u' = a$.

The first postulate is an existence postulate which shows that the system is not empty. So if one does not object to vacuousness, this postulate is dispensable. The second postulate requires that the K-rule of combination / shall be K-closed: it is a K-*closing* postulate. Postulate **3** is the principle of double negation. Postulate **4** states a relation between a, b, a' such that the empty class can be defined. The fifth postulate is a proposition of the law of distributivity for the K-rule of combination /.

Sheffer gives a proof of the equivalence of the set to Huntington's first postulate set of 1904. Such a proof cannot be carried out here; it will suffice to say that a set is shown to be equivalent to another by defining all the primitive ideas and rules of combination of the other and by proving its postulates as theorems and vice versa. This is what Sheffer does whilst demonstrating that there is a reciprocal implication between set **1–5** and Huntington's set.

Let us say, in concluding, that the development of Boole's logical calculus, as a pure formal process, has been achieved by the axiomatisation of Boolean algebra. The three sets of postulates for Boolean algebra which have been

---

[9] As a logical constant, Sheffer names the stroke '/', *rejection* 'for, if K is the class of all propositions of a given logical type, then whenever p and q are two propositions of this type, p/q may be interpreted as the proposition *neither p nor q*; in other words, / has the properties of the logical constant *neither–nor* (Sheffer 1913, p. 487).

described above have shown that the same logical system may be set up in several different ways. The system itself is abstract in the sense that it is a purely deductive theory, independent of the interpretations which may be given to the symbols. The logical concepts of the system are meaningful, but their meanings are not derived from experience, and the ways in which these concepts enter into combinations do not depend on the empirical world. Hence, the logical syntax of the system is constituted by the meaningful combinations and relations of the symbols.

However, the structure of the logical system may be applied to classes, to propositional forms, to several mathematical domains such as geometry, set theory, etc. But, as a logical syntax, Boolean algebra is an abstract system, of which its interpretation is irrelevant to the systematic structure. Thus, the logic of the system cannot be defined by the interpretation of the algebra. From this we can infer a fundamental principle: a logical system is syntactically independent of interpretation. It is, in other words, purely formal. Then arises this question: what is really the meaning of the K-elements of the sets and the rules of combination? They do not have meaning at all, inasmuch as the syntax of the system is not confined to any meaning in any locus whatsoever.

It should be noticed that all the sets of postulates for Boolean algebra use the words of ordinary language such as 'for all', 'there are at least', 'if...then', and thereby require a logical system involving a theory of quantification and rules for derivation. In other words, if one locus of Boolean algebra is a calculus of classes, then there must be a more fundamental theory about the inference of propositions from other propositions. Such a more fundamental theory is now provided by Frege's *Begriffsschrift*.

Chapter 5

An Account of the First Two Parts of Frege's *Begriffsschrift*

> *Cum deus calculat, fit mundus.*
> G. W. Leibniz

Introduction

Although Frege and Boole both referred to mathematics, this common reference should not conceal a difference of conception of the relation between logic and mathematics. They were aware that there is no sharp boundary between them. But, unlike Boole, Frege regarded logic as more fundamental than mathematics, and distinguished himself by introducing, for the first time, logic in mathematics. In the *Begriffsschrift* published in 1879, Frege constructed a logical language, that is, as the title of the booklet suggests, 'a formula language, modelled upon that of arithmetic, for pure thought.' He did so precisely because he needed such a logical tool, so as to draw mathematical inferences and definitions in a more rigorous way than could be done with ordinary language, and thereby to lay down the foundations of arithmetic as an extended logic.

In this chapter, I shall point out the particularity of Frege's work, which constitutes one of the two research programmes from which modern mathematical logic springs. I will sketch first the project of *Begriffsschrift*, which has an ideal of rigorous proof, and discuss how the concept of formal proof was conceived by Leibniz. Then I will give an account of the two first parts of

Begriffsschrift, which encompass two major innovations: the device of a 'language of pure thought' capable of representing perspicuously the conceptual content of a judgement; and a rigorously axiomatized presentation of propositional logic. Finally, I will discuss how far *Begriffsschrift* deserves its current predominant place in the history of logic.

But, first, I shall briefly inquire into the concept of logic, so as to elucidate how it is understood by Frege and Boole.

5.1 What is Logic About?

It is at about the time of the two-volume *Grundgesetze der Arithmetik* (*Basic Laws of Arithmetic*, 1893 and 1903) that Frege spelled out his conception of logic. In the book, he stressed the universal validity of the laws of thought, and the distinction between logic and psychological considerations. His concern was with the difference between logical and psychological laws and with the notions of objectivity and truth.

I shall sketch two aspects in which Frege's conception of logic may be apprehended: the view of logic as universal laws applicable to every science, and the principle of distinction between logic and psychology. Then, in the light of the epistemological framework whereby Frege approached this distinction, I shall show that Boole shares much the same conception of logic.

5.1.1 Logic as Universal

Frege regards logic as a science with laws, which are universal and applicable to every science. It determines the nature and conditions of valid reasoning, which scientists employ in attempting to introduce order into their disciplines. He says:

> It will be granted by all at the outset that the laws of logic ought to be guiding principles for thought in the attainment of truth... In one sense a law asserts what is; in the other it prescribes what ought to be. Only in the latter sense can the laws of logic be called 'laws of thought'... only if we mean to assert that they are the most general laws, which prescribe universally the way in which one ought to think if one is to think at all. (Frege 1893, I, XV)

Thus, the logical laws are laws of thought only in the sense in which they prescribe what ought to be and not what is. They are meant to be applicable to all scientific reasoning. Hence, logic becomes a standard in the sense that the laws which regulate this particular kind of reasoning are normative for any other kind of science that involves reasoning as a means of justifying its truth. But how does logic occupy this very central position amidst all other sciences? By its unqualified universality.

Weiner characterises the universality of logic as follows: 'the justification of a truth that can be inferred from the logical laws is not dependent of any particular facts. On the other hand, since the variables range over the entire universe, including the realm of natural sciences, Frege's logical laws can be used in inferences concerning any objects that are the subject of some special science' (Weiner 1990, p. 70). Thus, the structure of Frege's logical language justifies its universal applicability in inferences occurring in any science whatsoever. This structure is represented by the logical constants such as negation and the conditional signs, and by the variables. The quantifiers that bind the variables capture multiple generality, and permit generalisation over all objects. This syntactical universal language does not restrict itself to any particular branch of science, and so its truths are 'maximally general truths' concerned with concepts and relations of all science. As Ricketts says, 'purely logical laws then set forth generalizations that, not mentioning any objects or concepts investigated by the special sciences, do not distinguish among these. It is in this sense that logic, on the universalist conception, is the maximally general science' (Ricketts 1996, p. 123).

5.1.2 Psychologism versus Logic

There now arises the question as to how logic is concerned with truth. The attempt to answer this question prompts Frege to state a principle: 'always to separate sharply the psychological from the logical, the subjective from the objective' (Frege 1884, p. X). He does so whilst explaining what he means by 'being true' and 'being taken to be true':

> Being true is different from being taken to be true, whether by one or many or everybody, and in no case is to be reduced to it. There is no contradiction in something's being true which everybody takes to be false. I understand by 'laws of logic' not psychological laws of takings-to-be-true, but laws of truth.... If being true is thus independent of being acknowledged by somebody or other, then the laws of truth are not psychological laws: they are boundary stones set in an eternal foundation, which our thought can overflow, but never displace. (Frege 1893, I, XVI)

Frege prises apart logical laws and psychological ones. Psychological laws are subjective and dependent upon what the subject holds to be true, whereas the laws of logic are objective and independent of what is held as true by anyone.

In psychology, the laws of judgement are the same type as the laws of nature governing events in the external world. By contrast, in logic, the laws of thought are 'laws of valid inference' giving a proof of truth. They are not conceived as laws that describe how one comes to hold something as true. The laws of valid inference require that the truth of a thought rests upon previously assumed thoughts, that if the premises are true, then the following thoughts must also be true. They are concerned with the justification of thoughts, and tell us how one should think by giving a normative account of the correct use of thinking. Thus, logic is a regulative science. As Frege puts it,

> To make a judgement because we are aware of other truths as providing a justification for it is known as inferring. There are laws for this kind of justification, and to establish these laws of correct inference is the goal of logic. (Frege 1879/91, p. 3)

On the other hand, psychology is not concerned with justification of judgements, but rather with their 'causes' that have no inherent relation to truth. As Frege sees it, 'the causes which merely give rise to acts of judgement do so in accordance with psychological laws; they are just as capable of leading to error as of leading to truth; they have no inherent relation to truth whatsoever; they know nothing of the opposition of true and false' (Frege 1879/91, p. 2).

In Frege's eyes, psychologism as 'the corrupting incursion of psychology into logic' (Frege1893, XIV p. 12) has two varieties: (i) the reduction of truth to an individual's taking something to be true (Frege 1893, XV p. 13), and (ii) the belief in empirical evidence as basis for philosophy (Frege 1879/91, p. 2).[1] It erects an 'epistemological obstacle' which logic must surmount in order to reach the realm of objective knowledge. For him,

> It is the business of the logician to conduct an unceasing struggle against psychology and those parts of language and grammar which fail to give untrammelled expression to what is logical. He does not have to answer the question: how does thinking normally take place in human beings? What course does it *naturally* follow in the human mind? (Frege 1879/91, pp. 6–7)

Frege takes it that in order to eliminate psychologism from logic one must espouse realism or Platonism[2]. Indeed, in the late paper *Thoughts* (1918–19) he describes thoughts as neither things in the external world nor ideas. He

---

[1] These two varieties of psychologism are also pointed out in *The Foundations of Arithmetic* (1884) where Frege repels the invasion of psychology of mathematics by answering negatively these two questions: (i) is number something subjective? (p. 33); and (ii) are the laws of arithmetic inductive truths? (p. 12).

[2] There are different accounts of what Frege's Platonism amounts to. Thus, there is a philosophical dispute which flows from the enquiry into the question whether Frege is an

claims instead the existence of a 'third realm' in addition to that of the external world and the realm of subjective ideas, such that thoughts are abstract objects that exist independently of us in this realm.

5.1.3 Boole and Frege on Logic

The epistemological framework in which Frege poses the problem of psychologism seems to include Boole among those who are criticised. For Boole's references to 'mental acts' have prompted many authors to label his view of logic as a kind of psychologism. For instance, in *Frege: An Introduction to His Philosophy*, Currie argues that 'Boole's ideas on the philosophy of logic exemplify some aspects of psychologism' (pp. 15–16). As for Baker and Hacker, in *Frege: Logical Excavations*, they hold that 'Boole had taken it for granted that the laws of logic are in some sense laws of psychology' (Baker and Hacker 1984, p. 41).

But, if the two varieties of psychologism which Frege criticises are related to subjectivity, that is, the products of the individual mind and the use of inductive methods, then the characterisation of Boole's logical conception as psychologism does not hold water. Boole conceives the laws of logic as the same for all of us. This does not by itself show that they are not psychological. But he also holds that logic is independent of the thinking subject, and this does imply a non-psychological conception of logic. In *The Mathematical Analysis of Logic*, he heralds clearly that

> It will not be necessary that we should here enter into the analysis of that mental operation which we have represented by the elective symbol.... Our present concern is rather with the laws of combination and of succession, by which its results are governed, and of these it will suffice to note the following. The result of an act of election is independent of the grouping or classification of the subject. (Boole 1847, p. 16)

Accordingly, Boole aims to inquire into the fundamental laws of those operations through which reasoning is carried out, so as to express those laws in the symbolic language of logical calculus, and thus to demonstrate that logic is the science of reasoning. As he puts it, 'logic while it is the science of reasoning in general is in a more especial sense the science of reasoning by signs. It investigates the forms and the expressions to which correct reasoning may be reduced

---

epistemological or ontological Platonist. I shall refer the reader to Dummett, "Platonism", in *Truth and Other Enigmas*, (pp. 202–214); Joan Weiner, *Frege in Perspective* (pp. 176–224); Sluga, *Gottlob Frege*, (pp. 100–107); and Baker and Hacker, *Frege: Logical Excavations*, (pp. 59–62).

and the laws upon which it is founded' (Boole 1848a, p. 1). Hence Boole is not concerned with the question of 'how people actually think'. He wants rather to give a normative explanation of the *correct* use of reasoning. There is no trace of subjectivism in this investigation of the actual laws governing thought. Kneale corroborates this view when he writes that 'although Boole called his most ambitious work on logic *The Laws of Thought* and sometimes wrote as though he supposed himself to be investigating the constitution of the human intellect, it is clear that his algebra has nothing to do with thought processes. In each of the interpretations which we call logical it is concerned with relations between entities that are entirely non-mental' (Kneale 1962, p. 738).

There is also no trace of Mill's psychologism in Boole's logical investigations. Indeed, in *The Laws of Thought*, he disqualifies the use of inductive methods when he says

> The knowledge of the laws of mind does not require as its basis any extensive collection of observations. The general truth is seen in the particular instance, and it is not confirmed by the repetition of instances... A general truth in Logic... is made manifest in all its generality by reflection upon a single instance of its application. And this is both an evidence that the particular principle of formula in question is founded upon some general law or laws of the mind, and an illustration of the doctrine that the perception of such general truths is not derived from an induction from many instances, but is involved in the clear apprehension of a single instance. (Boole 1854, p. 4)

Certainly, Boole is here aiming at Mill, to whom he explicitly refers in a letter addressed to John Penrose on the 13th March 1855:

> On the other hand writers like Mill seem to me to be equally in error who maintain openly or by implication that propositions are unmeaning unless their terms relate to the distinctly conceived objects of individual experience.... It seems to be a law of human reason that we can in various instances affirm propositions without absolute certainty of their truth, respecting things which we can only picture or represent to ourselves as the limits of an indefinite process of abstraction. Nearly all if not all scientific truths are of this kind. (Boole 1855, p. 200)

Thus, Frege's arguments against psychologism in Mill's sense do not apply to Boole. On the contrary, Frege and Boole both criticise empiricism, which is a variety of psychologism which regards the laws of logic as obtained by induction from observations. Moreover, it is significant that whilst levelling his criticism at psychologism, Frege never mentions Boole as amongst those who advocates it. Knowing how polemical he is, it may be supposed that Frege is conscious of the resemblance between his logical conception and Boole's.

Indeed, in a fragment of Boole's manuscripts, there is the following view of logic similar to Frege's universalist conception of logic:

> The phenomenal study of things belongs to physics or to psychology, the inquiry into their absolute nature if indeed such an inquiry be possible is the business of metaphysics. Logic has other objects and is concerned with other relations—but as the relations with which it is concerned are universal (for all existing things can be contemplated under the notion of class or kind) Logic stands related to all other sciences. (Grattan-Guinness 1997, p. xlviii)

Regardless of everything that may particularise their two systems, Boole and Frege form the two sides of mathematical logic as a science. As Kneale puts it, 'Boole and Frege, like Leibniz before them, presented logic as a system of principles which allow for valid inference in all kinds of subject-matter...' (Kneale 1962, p. 739). Boole regards logic as having a subject matter which determines what follows necessarily from a proposition. Thus he says, 'when one or more propositions are given and from these we can infer the truth of some other proposition not identical with the given ones such a conclusion is obtained by a process of reasoning' (Boole 1848a, p. 2).

Furthermore, it should be pointed out that Boole even holds a surprising epistemological view close to Frege's conception of science. He writes,

> It may, perhaps, be permitted to the mind to attain a knowledge of the laws to which it is itself subject, without its being also given to it to understand their ground and origin.... Such knowledge is, indeed, unnecessary for the ends of science, which properly concerns itself with what is, and seeks not for grounds of preference or reasons of appointment.... *It is to be remembered that it is the business of science not to create laws, but to discover them.* (Boole 1854, p. 11)

As a scientist, Boole does not investigate the ground or origin of knowledge which is likely to lead only to metaphysical discussion. He rather seeks to discover the laws inherent in 'what is' and to exhibit them, so as to give their objective and universal characteristic. In his eyes, 'the object of science, properly so called, is the knowledge of laws and relations' (Boole 1854, p. 39). This view leads him to state an epistemological principle: the laws of reasoning must be independent of metaphysical theories of the nature of the mind. These laws have a real existence and 'contain an element of truth which no ulterior criticism upon the nature, or even upon the reality, of the mind's operations, can essentially affect' (Boole 1854, p. 40).

In effect, for Boole mathematical and logical propositions are necessary truths but, due to the imperfection of the senses, escape exact verifiability. He inquires about what these propositions really are. Are they merely a collection of experiences or does the mind supply some connecting principle of its own? For him, neither individual objects of experience, nor the mental images which they suggest, can give the correct answer. Mathematical and logical truths are, in some sense, out there to be discovered. Thus, Boole seems to hold a Platonist or realist position. He says,

> Although the perfect triangle, or square, or circle, exists not in nature, eludes all our powers of *representative* conception, and is presented to us in thought only, as the limit of an indefinite process of abstraction, yet, by a wonderful faculty of the understanding, it may be made the subject of propositions which are *absolutely* true. The domain of reason is thus revealed to us as larger than that of imagination. (Boole 1854, p. 405)

This position, which indicates that truths about mathematical objects are eternally true and are true independent of us, is exactly what Frege holds when he says,

> Numbers do not undergo change, for the theorems of arithmetic embody eternal truths. We can say, therefore, that these objects are outside time; and from this follows that they are not subjective percepts or ideas... (Frege 1895b p. 230)

5.2 The Project of *Begriffsschrift*

Although Frege was well acquainted with the formula language of mathematics, he makes clear that the modelling of logic upon the formula language of arithmetic, at which he hints in the title of his pamphlet, 'refers more to the fundamental ideas than to the detailed structure' (Frege 1879, p. 104). So he keeps logical symbols distinguished from those of arithmetic, so as to avoid confusing them. In his eyes, logic is both a *lingua characterica* and a *calculus ratiocinator* in the Leibnizian sense; hence logic is adequate for expressing mathematical inference, and not a mere calculus restricted to abstract logic. Unlike Boole, it is not in question for Frege to establish an 'artificial similarity' between the notations of arithmetic and logic, but to introduce logic into

An Account of the First Two Parts of Frege's *Begriffsschrift*   145

mathematics through the logical reconstruction of arithmetic. This leads Frege to design the *Begriffsschrift*.

5.2.1 An Ideal of Rigour in The Proof

Frege believes that mathematics is derivable from logic, and he wants to employ a logical tool to prove it. But, he realises that he cannot succeed without developing a 'formula language of pure thought', for ordinary language is not suitable to represent proofs. Hence, it is the ideal of a perfect rigour in mathematical procedure that gives rise to the project of *Begriffsschrift*.

Indeed, Frege strives 'to reduce the concept of ordering-in-a-sequence to that of *logical* ordering, in order to advance from here to the concept of number' (Frege 1879, p. 104). There follows the project of *Begriffsschrift*:

> So that something intuitive could not squeeze in unnoticed here, it was most important to keep the chain of reasoning free of gaps. As I endeavoured to fulfil this requirement most rigorously, I found an obstacle in the inadequacy of language; despite all the unwieldiness of the expressions, the more complex the relations became, the less precision—which my purpose required—could be obtained. From this deficiency arose the idea of the 'conceptual notation' presented here. Thus, its chief purpose should be to test in the most reliable manner the validity of a chain of reasoning and expose each presupposition which tend to creep in unnoticed, so that its source can be investigated. (Frege 1879, p.104)

Frege aims to exhibit a chain of inference in which there are no gaps. It is not only required to make explicit the mathematical principles which provide the content of mathematics, but also the underlying logical principles which ensure its formal structure.

The logical connections between mathematical formulae cannot be expressed in ordinary language, which is imprecise and ambiguous. But there is symbolism for mathematical procedure which can exhibit the validity of mathematical definitions and inferences. Hence, in order to achieve an ideal of rigour in proof, Frege builds up a formula language which is more adequate than ordinary language.

Accordingly, the necessity of a formula language stems from the inadequacy of ordinary language as a tool for the aim of representing perspicuously logical inferences involved in mathematical proofs. Frege then constructs a 'language of pure thought', that is, a system of conceptual notation with all the rigour it needed to ensure that mathematical inferences are protected from the imprecisions and the ambiguities of ordinary language. He presents it as the support and the object of a logical calculus, which would be then performed independently of the meaning of the symbols.

Frege has already discovered that Aristotelian logic, which analyses propositions in terms of subject and predicate, is not also an appropriate instrument to represent all the logical principles that enter into the formal structure of mathematical inferences. Instead, he sets out to analyse propositions into function and argument, which then allows him to carry out the validity of inferences concerning with multiple generality that occur in mathematics. As a result, he performs a revolution in logic.

The project of *Begriffsschrift* is then the construction of a notation for representing inferences and the setting up of a formal system for rigorously testing their validity.

5.2.2 Leibniz and The Concept of Formal Proof

In the preface of the *Begriffsschrift*, Frege speaks of the attempts made by his predecessors like Leibniz who, had had an idea of a universal characteristic, called *calculus philosophicus* or *ratiocinator*, that is, of an adequate system of notation which represents by appropriate symbols concepts and logical operations, and submits them to a formal algorithm.

I shall briefly describe the intellectual ambience in which the concept of formal proof emerged at the time of Leibniz, who posed the 'bare preliminaries' of the formalisation of logic, which Boole then developed elegantly in his logical investigations.

In the seventeenth-century intellectual revolution, which laid down the foundations for the modern scientific epoch, it was increasingly common to discredit logic. Descartes was particularly famous for vehemently opposing Aristotelian-scholastic deduction. For him, the criterion of truth being clarity and distinctness, it should not be in question for reason to rely upon the traditional logical rules. The mnemonic verses such as bArbArA and cElArEnt are useless: they are pure psittacism which blunts our reason and should then be booted out of philosophy to rhetoric. Truth can solely be grasped by what he called *inspectio mentis*, that is, the intuitive or psychological perception of a clear and distinct idea. At the bottom of *Regula X*, he wrote:

> It may perhaps strike some with surprise that here, where we are discussing how to improve our power of deducing one truth from another, we have omitted all the precepts of the dialecticians, by which they think to control the human reason. They prescribe certain formulae of argument, which lead to a conclusion with such necessity that if the reason commits itself to their trust, even though

An Account of the First Two Parts of Frege's *Begriffsschrift* 147

> it slackens its interest and no longer pays a heedful and close attention to the very proposition inferred, it can nevertheless at the same time come to a conclusion by virtue of the form of the argument alone.

Since Descartes wanted to pay 'a heedful and close attention to the very proposition inferred', so as to keep his mind attentive whilst examining the truth, he rejected the formulae of argument prescribed by the logicians called 'dialecticians', and then added:

> It may appear still more evident that this style of argument contributes nothing at all to the discovery of the truth, we must note that the Dialecticians are unable to devise any syllogism which has a true conclusion, unless they have first secured the material out of which to construct it, i.e. unless they have already ascertained the very truth which is deduced in that syllogism. Whence it is clear that from a formula of this kind they can gather nothing that is new, and hence the ordinary dialectic is quite valueless for those who desire to investigate the truth of things. Its only possible use is to serve to explain at times more easily to others the truths we have already ascertained; hence it should be transferred from Philosophy to Rhetoric. (Descartes 1955, pp. 32–33)

This depreciation of the scholastic teachings is symptomatic of the ignorance of Descartes of what a proof is. As Hacking tells us, 'Descartes thought that truth conditions have nothing to do with demonstration' (Hacking 1973, p. 1). However, what the seventeenth century called 'geometers' method' is a method of proof. Geometry is indeed a deductive science based upon definitions and axioms which allow the inference of theorems.

Leibniz, who knew what proof is, recognised the importance of geometry as a formal procedure. Hence by attempting to restore to logic its status of rigorous science, he pointed out that

> Logic admits of demonstration as much as geometry does, and geometers' logic that is, the methods of argument which Euclid explained and established through his treatment of proportions, can be regarded as an extension or particular application of general logic. (Leibniz 1765, IV, II, 13, p. 371)

He contrasted the Cartesian psychological criterion of truth with the method of geometers, which provides proofs, and thereby demonstrates their conclusions to all human beings. Consequently, Leibniz assessed logic differently from

Descartes and the majority of the thinkers of his time. In the *New Essays*, in responding to Locke who, wanted to get rid of logic, he corrected the current unbridled attitudes against logic as follows:

> I hold that the invention of the syllogistic form is one of the finest, and indeed one of the most important, to have been made by the human mind. It is a kind of universal mathematics whose importance is too little known. It can be said to include an *art of infallibility*, provided that one knows how to use it and gets the chance to do so- which sometimes one does not. But it must be grasped that by 'formal argument' I mean not only the scholastic manner of arguing which they use in the colleges, but also any reasoning in which the conclusion is reached by virtue of the form, with no need for anything to be added. So: a sorites, some other sequence of syllogisms in which repetition is avoided, even a well drawn-up statement of accounts, an algebraic calculation, an infinitesimal analysis- I shall count all of these as formal arguments, more or less, because in each of them the form of reasoning has been demonstrated in advance so that one is sure of not going wrong with it. (Leibniz 1765, IV, XVII, 4, p. 479)

This passage itself sums up the view of Leibniz on traditional logic. On the one hand it reveals the admiration he had for the syllogistic forms of reasoning. But on the other hand it shows clearly that this admiration concerns the formal proof which underlies these procedures. In opposition to Descartes and Locke, who condemned formalism, Leibniz knew what a proof is, that is, 'a sequence of sentences beginning with identities and proceeding by finite number of steps of logic and rules of definitional substitution to the theorem proved' (Hacking 1973, p. 4). Hence, he admired syllogistic theory in that the inferences are valid in virtue of its form, not its content. As Hacking sees it,

> The merit of the old system was that it gave us some understanding of the nature and interconnection of truths. The demerit was the inadequacy of the implied methodology of doing physics by deduction. So Leibniz grafted a new methodology on the old theory of demonstration.... It is turned into the theory of formal proof. In the old tradition only universal propositions are subject to demonstration. In the new practice, only what we now call pure mathematics fits this model. (Hacking 1973, p. 14)

Now, the new methodology that Leibniz grafted on the old theory of demonstration is envisaged as a project of formulating a universal characteristic,

> Which renders truth stable, visible and irresistible, so to speak, as on mechanical basis.... Algebra, which we rightly hold in such esteem, is only a part of this general device. Yet algebra accomplished this much- that we cannot err even if we wish and that truth can be grasped as if pictured on paper with the aid of a machine. I have come to understand that everything of this kind which algebra proves is due only to a higher science, which I now usually call a *combinatorial characteristic*. (Hacking 1973, p. 5)

This combinatorial characteristic provides a kind of thread that could guide us infallibly into the labyrinth of proofs featuring any demonstration. It is a direct result of the marriage between mathematics and logic. It turns resolutely towards the future in that it anticipates modern logic by laying down the foundation upon which modern mathematical logic is built.

Leibniz first designed his project in his early work in which he heralded the undertaking to set up a system of written symbols separated from speech, and directly expressing the fact.[3] Then, he discussed it later in the *New Essays*, where he aimed to overcome the difficulties relating to 'general truths' that are almost incomprehensible because 'conceived and expressed in words'. He suggested replacing the spoken words with a 'Universal Symbolism—a very popular one', that is, a system of 'little diagrams' independent of ordinary language in which the 'little diagrams' and their combination stand directly for things and not words, in such a way that each people may express and understand them on their own (Leibniz 1765, VI, II, 13, p. 399). It is a symbolical system, or an ideography language in which the words and their combinations are depicted in characters or signs to ensure the possibility of fixing firmly to the paper diverse sequences of reasoning, so as to enable one to grasp at a glance their syntactical structure. Such a pictorial symbolism would literally 'speak to the eyes', and would sustain our steps into the labyrinth of proofs like 'the little wheeled device which keeps toddlers from falling down' (Leibniz 1765, IV, II, 13, pp. 371–372).

Leibniz built up his system of logic in such a way as to make 'a chain of reasoning which will represent all the argumentation even of an orator. It will have been stripped of its ornamentation and reduced to the bare bones of "logical form"' (Leibniz 1765, IV, XVII, 4, p. 481). He preferred the numbers and the symbols of algebra for expressing his characteristic rather than Chinese or Egyptian hieroglyphs, because the former allows us to carry out reasoning and demonstrations by a calculus analogous to those of arithmetic and algebra,

---

[3] In his 'Dissertation on Combinatorial *Art*' written in 1666, Leibniz in studying the combinatorial method, which is based upon the decomposition of ideas in simple elements, claimed '*Notas quam maxime naturales*'.

whereas the latter serve to represent ideas or facts. Hence Leibniz realised that the universal characteristic was a symbolic language that could be treated as a calculus of reasoning, like an algebra. Thus, the characters represent the 'alphabet of human thought' within which we could define all our concepts through the operations of negation, conjunction, and disjunction. In other words, they are primitives concepts from which complex concepts could be built by means of rules of combination. Leibniz called this process of building complex concepts 'the art of combination', which is a calculus of reasoning.

It follows that the logical project of Leibniz may be seen as mathematics under the two aspects of *lingua characterica* and of *calculus ratiocinator*, which both laid down the foundation upon which modern mathematical logic is built. He modelled his system of logic upon mathematics as a universal formula language, because what makes mathematics successful in its demonstrations is the fact that, as a deductive science, it proceeds apart from the contents of its propositions. It then appears that mathematics and logic both aim at an ideal of formal proof which should connect them inextricably. But, as opposed to modern conceptions of mathematical logic, Leibniz failed to realize that the relations involved in his combinatorial characteristic must be analysed. He took for granted the truth of the primitive concepts since they were to be given by means of 'the alphabet of human thought'. In truth this seems to be a mistake since, as has been seen, (see section **4.3**) the primitive concepts of an axiomatic system are not given; they are, to some extent at least, arbitrarily singled out.

In addition, Leibniz did not really achieve his logical project, which came to us in a fragmentary form.[4] But in the nineteenth century, following the way paved by Leibniz, Boole performed a *calculus ratiocinator* by clothing logic in the dress of algebraic symbolism, and thereby carried out a formal proof which a scientific theory of deductive reasoning needed. Thus, it can be said that Leibniz and Boole, who both knew the importance of formal proof in a system of truths, began what *Begriffsschrift* subsequently developed more systematically: a Universal Characteristic, which represents perspicuously logical relations.

In this connection Russell holds:

> Mathematical logic...is mathematical in two different senses: it is itself a branch of mathematics, and it is the logic which is specially applicable to other more traditional branches of mathematics.

---

[4]It is Couturat and Russell's interpretation of his logic that makes it possible to reconstruct now Leibniz's latent idea of a mathematical logic. This interpretation can be found in Couturat:*La Logique de Leibniz*, 1985; and Russell: *A Critical Exposition of The Philosophy of Leibniz*, 1900.

An Account of the First Two Parts of Frege's *Begriffsschrift*   151

> Historically it began as *merely* a branch of mathematics; its special applicability to other branches is more recent development. In both respects, it is the fulfilment of a hope which Leibniz cherished throughout his life, and pursued with all the ardour of his amazing intellectual energy. (Russell 1922, p. 49)

Although it seems to be sketchy, it may be plausible to infer from Russell's analysis that Boole's logical system is the realisation of the first aspect of the logical project of Leibniz, which makes it a 'branch of mathematics', whereas Frege's logical system is the achievement of the second aspect of characteristic language 'applicable to other more traditional branches of mathematics'.

5.3 A Formula-Language For Pure Thought

Frege attempts 'a fresh approach to the Leibnizian idea of a *lingua characterica*'. As a mathematician, he realises that the condition for introducing correct reasoning in mathematics requires a symbolic notation which represents the logical structure of reasoning of any kind. Hence, he builds up a system of signs written with special symbols 'for pure thought', expressing 'conceptual content.' In this formula-language, the symbols and their combinations must be precisely defined in such a way that the rules of deduction assure the validity of the reasoning. As a result, Frege carries out for the first time the Leibnizian idea of a *lingua characterica*.

5.3.1 Frege's Symbolic Notation

Frege divides all the symbols he employs into *'those which one can take to signify various things* and *those which have a completely fixed sense'* (Frege 1879, p. 111). The first are the letters which serve for the expression of generality. But, although we can give to a letter any sense we wish, Frege insists that we must retain the same sense throughout the context in which the letter occurs. The second serve for the expression of negation, conditionality and identity. As in mathematics, the terms 'variables'[5] and 'constants' may be respectively used to designate these two kinds of symbols. But these terms are

---

[5] Frege prefers to call the symbols 'italicised letters' instead of using the expression 'variable', which he regards as misleading in his paper, 'Logical Defects in Mathematics' (see Frege 1898/99, pp. 159–166). Indeed, as Quine says it, 'a variable is not best thought of as somehow varying through time, and causing the sentence in which it occurs to vary with it. Neither is it to be thought of as an unknown quantity, discoverable by solving equations' (Quine 1974, p. 122).

best understood only when Frege introduces his semantic theory.[6] In the first chapter of *Begriffsschrift* Frege lets the capital letters '*A*' and '*B*' abbreviate the propositions. But he does not follow this practice in the second chapter where he takes small case letters '*a*', '*b*' and so on as abbreviations for the propositions.

Since Frege's notation is concerned with 'conceptual content', he distinguishes 'judgment' from its mere 'content' and employs two signs to capture the distinction. I shall give the definition of the symbols employed in Frege's conceptual notation.

5.3.2 The Content-Stoke

The content of a judgment is regarded as a 'combination of ideas'; that is, a complete sentence and its sign is 'combined...into a whole' by a horizontal stroke prefixed to the sign:

The long horizontal line to the left of the proposition is called the 'content stroke', that is, a 'mere combination of ideas, [7] of which the writer does not state whether or not he acknowledges its truth' (Frege 1879, p. 111). The complete symbol can be read as 'the circumstance that' or the 'proposition that'.[8] This paraphrase intends to highlight the fact that the assertible-content does not have an assertoric force.

5.3.3 The Judgment-Stroke

If the given content is asserted for expressing a judgment, that is, to acknowledge its truth, then the 'judgment stroke' is added to the left end of the content stroke:

---

[6]For, as Church defines them, a variable is a symbol which has a certain non-empty range of possible values; and a constant is a proper name having a denotation (Church 1956, p. 9). A variable for which sentences expressing propositions may be substituted is called a propositional variable. The range of a propositional variable is the two truth-values.

[7]Instead of combination of 'ideas', the reader can read 'thought', for Frege writes that he now simply says "*Gedanke*" (see Van Heijenoort 1970, footnote 6).

[8]Instead of 'circumstance' and 'proposition' Frege would simply say 'thought' (see Van Heijenoort 1970, footnote 9).

An Account of the First Two Parts of Frege's *Begriffsschrift* 153

For Frege '...the horizontal stroke...ties the symbols which follow it into a whole; and the assertion, which is expressed by means of the vertical stroke at the left end of the horizontal one, relates to this whole' (Frege 1879, p. 112). Thus, the judgment stroke expresses the 'assertion' of the whole formed by the sign for the content and the content stroke.[9] The composite prefix may be construed as 'the common predicate of all judgments', in so far as any proposition can be given in the form '...is a fact', e.g., 'The violent death of Archimedes at the capture of Syracuse is a fact'. Here then 'the subject contains the whole content, and the predicate serves only to present this as a judgment.' Frege says that 'such a language would have only a single predicate for all judgments, namely "is a fact"'[10] (Frege 1879, p. 113). Frege introduces the judgment-stroke or the assertion sign in order to distinguish rigorously judgement from assertible-content within the formal notation for inference. He points out that 'the relation in a hypothetical is not one between judgements but between contents of possible judgement. But if I affirm that this relation holds, I then express a judgement' (Frege 1880/81, 11n).

5.3.4 The Negation-Stroke

The negation-stroke is placed vertically under the content-stroke dividing it into two parts. The part to the right of the negation stroke is the content stroke of the original content, the part to the left that of its negation. The whole expresses that the 'content does not occur' without expressing whether this thought is true.

---

[9] In *Principia Mathematica*, Whitehead and Russell employ the assertion sign and put it before their symbolic propositions. But the presence of the sign shows that 'there is no need of the distinction between real and apparent variables, nor of the primitive idea "assertion of a propositional function."' According to them, an asserted proposition of the form '⊢ fx' is taken as meaning '⊢ $(x)fx$.' (Whitehead and Russell 1927, vol. 1, p. XIII). In *Mathematical Logic*, Quine comes near enough to this interpretation to justify his retention of the notation. He employs the sign as a syntactical notation which plays the role of an abbreviation. He writes "⊢ (ϕ)" to mean that the closure of ϕ is a theorem (Quine 1961, p. 88). In *Introduction to Mathematical Logic*, Church uses the sign 'as a syntactical notation to express that a well-formed formula is a theorem. Thus "⊢ $p \supset p$" may be read as an abbreviation of "$p \supset p$ is a theorem"' The notation enables him to state a metatheorem about simultaneous substitution which he then uses as a *derived rule of inference* (Church 1956, p. 83). This has become standard notation.

[10] This view according to which the judgement stroke may be regarded as equivalent to the common predicate of all judgements, that is, '...is a fact' was repudiated by Frege in *On Sense and Reference* (1892), where he argues that a truth-value cannot be a part of thought but the reference of sentence (see Frege 1892a, p. 64).

Thus, negation attaches to the underside of the content stroke and is a mark of a possible content of judgment but not an act of judgment.

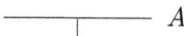

5.3.5 The Conditional-Stroke

Frege introduces the condition-stroke as follows: If A and B stand for possible contents of judgment, there are the four following possibilities:

A affirmed, B affirmed
A affirmed, B denied
A denied, B affirmed
A denied, B denied.

In this tabulation of the four possibilities, Frege, without giving any further justification, does not use the word 'true' and 'false' but rather 'affirmed' and 'denied'. In fact, in the *Begriffsschrift,* he does not employ truth-values. It can be, however, inferred from his explanation of the meaning of the conditional stroke that he is thinking of using the truth-table method. There follows a puzzling point, in so far as it is quite clear that in his *Foundations of Arithmetic* and the publications that come after, the conception of truth constitutes a determining factor in Frege's logical writings. Why does Frege move from 'denied' to 'false', from 'affirmed' to 'true'? Sluga conjectures that this change of terminology is accompanied by a new interest in the notion of objectivity, which is an essential part of the attempt to clarify the notion of truth (Sluga 1980, p. 111). Certainly, this explanation that links objectivity, thought, or sentence to truth would dovetail with Dummett's view that 'the root notion of truth is then that a sentence is true just in case, if uttered assertorically, it would have served to make a correct assertion' (Dummett 1991a, pp. 165–166).

The formula:

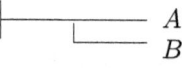

'stands for the judgment that the third of these possibilities does not occur, but one of the other three does.' In other words, 'it is denied that A denied and B affirmed' namely, the third possibility. For Frege, 'the vertical stroke which connects the two horizontal ones is the *conditional stroke*. The part of the upper horizontal stroke situated to the left of the conditional stroke is the content stroke for the meaning... of the symbol combination; to this is attached every

An Account of the First Two Parts of Frege's *Begriffsschrift*

symbol which is intended to relate to the content of the expression as a whole. The part of the horizontal stroke lying between A and the conditional stroke is the content stroke of A. The horizontal stroke to the left of B is the content stroke of B' (Frege 1879, p. 116). Frege holds that we can make a judgement expressed by the above formula without knowing whether A and B are to be affirmed or denied. As an example, he takes B to stand for 'the circumstance that the moon is in quadrature [with the sun]', and A 'the circumstance that it appears as a semicircle'. In this case, he claims that the formula can be translated with the aid of the connective 'if': 'if the moon is in quadrature [with the sun], it appears as a semicircle' (Frege 1879, pp. 115–116). Thus, Frege reintroduces the symbolical expression of the conditional that turned out to be the 'material' conditional, which was favoured by Philo of Megara in antiquity but had been cast away by logicians (except Boole). Frege notices that the third possibility 'not (B and not $-A$)' has some logical properties in common with the ordinary language expression 'if B then A'. But he also stresses that the symbol does not fully correspond to the conjunction 'if'. He says that 'the causal connection implicit in the word "if", however, is not expressed by our symbols, although a judgement of this kind can be made only on the basis of such a connection; for this connection is something general, but at this point we do not yet have an expression for generality' (Frege 1879, p. 116). This last comment indicates that Frege holds that the introduction of the notation for generality enables him to express perspicuously causal connections.

From the sign of conditionality, Frege draws the expression of a simpler more perspicuous rule of inference that is, the rule of *modus ponens* (see subsection **5.4.2**). When combined with the sign of negation, the sign of conditionality also enables him to define the other signs:

is translated by 'the case in which B is to be affirmed and the negation of A is to be denied does not occur'; or 'the possibility of affirming both A and B does not exist'. 'A and B exclude each other'. Hence, the following cases are obtained:

A affirmed, B denied
A denied, B affirmed
A denied, B denied.

156   An Account of the First Two Parts of Frege's *Begriffsschrift*

The negation of the whole content of this judgement gives the formula:

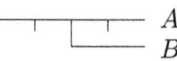

which is translated by 'the case in which A and B are both affirmed occurs', or 'Both A and B are facts': that is the conjunction. Hence, the following case is obtained:

A affirmed, B affirmed.

Frege could have used conjunction as primitive connective instead of the conditional stroke. The great advantage, however, of taking the conditional as primitive is the fact that it connects directly with *modus ponens* as a simpler rule of inference. As he puts it, 'instead of expressing "and" by means of the symbols for conditionality and negation, as is done here, we could, conversely, represent conditionality by means of a symbol for "and" and the symbol for negation' (Frege 1879, p.123). The formula:

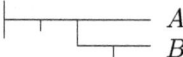

means 'the case in which A is denied and the negation of B is affirmed does not obtain', and is translated by 'A and B cannot both be denied': that is the disjunction. Hence the following cases are obtained:

A affirmed, B affirmed
A affirmed, B denied
A denied, B affirmed.

Another way of describing this case is as one where ' A or B' is affirmed, as long as one adopts the inclusive use of the expression 'A or B' in which the joint affirmation of A and B is not excluded; of course this was not Boole's view (see subsection **3.2.1**).

The Sign of Identity

Frege introduces a sign for identity of content by saying that the formula:

$$\vdash \quad (A \equiv B)$$

means 'the symbol A and the symbol B have the same conceptual content, so that we can always replace A by B and vice versa'. This needs considerable interpretation if misunderstanding is to be avoided. The now familiar interpretation of '$A \equiv B$' as 'A if and only if B' is not what Frege has in mind. Instead 'A' and 'B' here are individual variables and Frege is indeed thinking of cases which we might express as '$a = b$'.[11] Thus, unlike conditionality and negation, the judgement expressed by the formula is about names, and not contents. Frege says,

> Although symbols are usually only representatives of their contents so that each combination (of symbols usually) expresses only a relation between their contents- they at once appear in *propria persona* as soon as they are combined by the symbol for identity of content, for this signifies the circumstance that the two names have the same content. (Frege 1879, p. 124)

Since the same content may be given in different way, he argues that a symbol for identity of content is necessary. As an illustration of his treatment of identity as a relation between two names that have the same content, Frege uses an example from geometry. He gives the name 'A' which represents a fixed point on a circle, and the name 'B' which represents the point of intersection of the circle and a ray revolving about A. When the revolving ray is perpendicular to the diameter drawn from A, then 'A' and 'B' represent the very same point. Thus the names 'B' and 'A' has the same content. But, in this example, Frege does not consider the manner in which the content is determined (as the point on the circle) as an aspect of meaning.

This will be done in his '*Über Sinn und Bedeutung*', where he splits the notion of content into sense and reference. In this article, Frege changes his view that identity is a relation between names. He sets out to give an account of the sign of identity in a more satisfactory way than was done in the *Begriffsschrift*. In order to explain the fact that there are statements of the form '$a = b$' which 'contain very valuable extensions of our knowledge and cannot always be established *a priori*', Frege holds that the sign 'a' is not distinguished from

---

[11]In *The Basic Laws of Arithmetic*, Frege retains all the symbols of *Begriffsschrift* except '$\equiv$' which is replaced by the sign of equality.

the sign 'b' only as object. If we assume '$a = b$' to be true, then '$a = a$' and '$a = b$' can have different cognitive values only if 'a' is distinguished from 'b' not only in form but also in 'the manner in which it designates something'. 'A difference can arise only if the difference between the signs corresponds to a difference in the mode of presentation of that which is designated' (Frege 1892a, p. 57). Using again an example from geometry, he considers a, b, c, as the lines connecting the vertices of a triangle with the midpoints of the opposite sides. An elementary geometric statement tells us that the point of intersection of a and b is identical with that of b and c so that the statement 'the point of intersection of a and b is the same as the point of intersection of b and c' is true. So the expressions 'the point of intersection of a and b' and 'the point of intersection of b and c' are different designations for the same point. But these names 'likewise indicate the mode of presentation; and hence the statement contains actual knowledge' (Frege 1892a, p. 57). Thus, Frege brings forth, in addition to the designation which he calls the reference, something else 'wherein the mode of presentation is contained' and he calls this the sense of the sign (Frege 1892a, p. 57). Accordingly, he says that 'the reference of the expressions "the point of intersection of a and b" and "the point of intersection of b and c" would be the same, but not their senses' (Frege 1892a, p. 57).

Furthermore, the sign of identity allows the expression of definitions occurring in mathematical proofs. For Frege, in these formal definitions, the sentence does not say, 'the right side of the equation has the same content as the left side'; but, 'they are to have the same content'. This sentence is therefore not a judgement; and thus also not a judgement of the kind Kant called synthetic. What stands on the left side of the symbol of identity is the *definiendum* or the expression being defined, and on the right side the *definiens* or the expression supplying the definition. Regarding proofs, a definition has as a function to permit the replacement of the *definiendum* with the *definiens*, thus unfolding what was previously wrapped in the *definiendum*. According to Frege 'the only aim of such definitions is to bring about an extrinsic simplification by the establishment of an abbreviation. Besides, they serve to call special attention to a particular combination of symbols from the abundance of the possible ones and thereby obtain a firmer grasp [of it] for the imagination' (Frege 1879, p. 168).

5.3.6 Frege's New Theory of Judgement: Function / Argument

One of the most important innovations of the *Begriffsschrift* is Frege's new theory of judgement. It leads him to introduce, for the first time in the history of logic, the device of quantification for the analysis of general propositions.

An Account of the First Two Parts of Frege's *Begriffsschrift*

Frege himself believes that 'the replacement of the concepts of *subject* and *predicate* by *argument* and *function*, respectively will prove itself in the long run (Frege 1879, p. 107). In order to understand Frege's advance over Aristotelian and Boolean logic in this respect, it will help to begin by sketching how the theory of judgement stood before him.

Aristotelian logic played a fundamental role in the theory of judgement which was dominant before Frege. It was concerned with the evaluation of inferences whose validity was based on relations between terms. As a logic of general terms, it dealt with the various patterns of argument which could be represented by combining the expressions 'all', 'some', 'no' with terms. According to Aristotelian logic every judgement can be analysed as composed of a subject-term and a predicate-term.

But the analysis of judgements in terms of subject-term and predicate-term involved difficulties. Indeed, it concealed propositions of very different forms by assimilating singular propositions (e.g. 'Socrates is mortal') to universal propositions (e.g. 'All Greeks are mortal'). And since judgements are obtained by combining terms in various sharply delimited ways, it tolerated few forms of judgements. For instance, the analysis could not accommodate relational predicates. The proposition, 'Hydrogen is lighter than carbon dioxide' could not be analysed in this way. Moreover, the most important difficulty was its inability to deal with propositions involving multiple generality (e.g. 'Every son is the child of some father') which the medieval logicians tried to account for by means of their theory of supposition. An example of a valid inference for which Aristotelian logic could not provide involving such multiple generality was the following:

> Every man is an animal,
> Every head of a man is a head of an animal.

Yet propositions involving multiple generality are abundant in mathematics (e.g. 'Every natural number is exceeded by at least one natural number'). It follows that if one is to be able to carry out the logical analysis of mathematical inferences, then a satisfactory account of multiple general propositions is required.

Regarding these difficulties, Boolean logic had not gone beyond Aristotelian logic. Boole was indeed never able to advance much beyond the principles of Aristotelian logic because he lacked a proper theory of quantification. He retained the Aristotelian analysis of propositions in terms of subject-term and predicate-term, with the copula conceived as identity. For him, the analysis of a proposition consists of identifying the subject and the predicate as distinct terms within the proposition, which are somehow 'given' prior to the formation of the proposition. He assumed that judgement expresses identity of the denotation of its terms: Boole's conception of terms and their relation to judgement was basically the Aristotelian conception. Admittedly, although Boole did not deal himself with the logic of relative terms, it was carried out within his research programme (see subsection **2.1.2**).

However, the logical system which Frege develops in the *Begriffsschrift* overcomes the difficulties of Aristotelian and Boolean logic once and for all. He introduces a new theory of judgement which provides a more flexible tool for the analysis of propositions. Indeed, the function / argument analysis of propositions opens up an infinite space of logical forms. Functions can have several arguments, and they can be picked out of any level of the hierarchy of functions. Frege develops polyadic predicate logic i.e. a logic using predicates that take any number of arguments. Unlike Aristotelian and Boolean logic, polyadic predicate logic brings out the detailed structure of relational predicates and just includes monadic predicate logic; i.e. a logic using predicates that take one argument, as a simple case. The development of polyadic predicate logic is very important since a great number of valid inferences in mathematics and empirical science cannot be proved via the methods appropriate to either propositional or monadic predicate logic.

Frege writes

> A distinction between *subject* and *predicate* does not *occur* in my way of representing a judgement. To justify this I note that the contents of two judgements can differ in two ways: first, it may be the case that [all] the consequences which can be derived from the first judgement combined with certain others can also be derived from the second judgement combined with the same others; secondly this may not be the case. The two propositions, 'At Plataea the Greeks defeated the Persians' and 'At Plataea the Persians were defeated by the Greeks', differ in the first way.... Now I call the part of the content which is the *same* in both *conceptual content*. (Frege 1879, pp. 112–113)

Two judgements may differ in various ways which do not affect their contributions to the validity of any inferences in which they occur. Frege introduces what he calls, 'conceptual content', for what two such judgements have in common. Logic for him is the study of the objective contents of judgements. The grammatical difference between subject and predicate is of no relevance. Accordingly, Frege rejects the grammatical notions of subject and predicate and substitutes for it the mathematical notions of function and argument. He gives the formal description of function and argument as follows:

> If, in an expression (whose content need not be assertible), a simple or a complex symbol occurs in one or more places and we imagine it as replaceable by another [symbol] (but the same one each time), at all or some of these places, then we call the part of the expression that shows itself invariant [under such replacement] a function and the replaceable part its argument. (Frege 1879, p. 127)

In the *Begriffsschrift*, although Frege mentions that 'arguments' are symbols which denote an object (i.e. names), function and argument are called expressions, not what expressions represent or denote. Moreover, he makes it clear that the distinction between function and argument does not concern the conceptual content. As he emphasises, 'this distinction has nothing to do with the conceptual content, but only with our way of viewing it' (Frege 1879, p. 127).

Since Frege seems to be concerned with linguistic functions in sections 9 and 10 of *Begriffsschrift* in which he introduces the notion of function, let us consider the following sentence about the legendary outlaw of twelfth-century England as an illuminating explication of the notion,

'Prince John outlawed Robin Hood'

and replace the name 'Prince John' with the name 'King Richard'. In this way, its content is altered from a true sentence into a false sentence. The sentence may be considered as being built up of a constant component

'_____ outlawed Robin Hood'

and a replaceable symbol,

'Prince John'.

The name of 'Prince John' is replaceable by names which designate other persons in the same way as 'Prince John' designates Prince John. Now, in a sentence built up in this way, Frege considers the first fixed component as a function, and the second component as the argument of the function. Thus, the sentence

'Prince John outlawed Robin Hood'

is the result of completing the expression

'_____ outlawed Robin Hood'

with the name

'Prince John',

and the sentence

> 'King Richard outlawed Robin Hood'

is the result of completing the same expression with the name

> 'King Richard'.

In other words, the sentence 'Prince John outlawed Robin Hood' is the value of the function '_____ outlawed Robin Hood' for the argument 'Prince John', and 'King Richard outlawed Robin Hood' is the value of the same function for the argument 'King Richard'. Moreover, a sentence may be analysed in more than one way. Thus the sentence

> 'Prince John outlawed Robin Hood'

is not only the value of the function '_____ outlawed Robin Hood' for the argument 'Prince John'; it is also the value of the function 'Prince John outlawed _____' for the argument 'Robin Hood'. The functions '_____ outlawed Robin Hood' and 'Prince John outlawed _____' are unsaturated and each needs to be completed by a single name to turn them into a sentence; they are functions of an argument.

It should be made clear here that, in the *Begriffsschrift*, Frege views the sentence, 'hydrogen is lighter than carbon dioxide' itself as the value of the different functions for different arguments, just as is shown with the sentence 'Prince John outlawed Robin Hood'. Frege is concerned there with linguistic notions: sentences, the names they contain and the functions which map names as arguments onto sentences as values. It is only after the appearance of *The Foundations of Arithmetic* (1884) that Frege thinks of truth-value as the value of a function. In his paper, 'Function and Concept', he writes

> I now say: 'the value of our function is a truth-value', and distinguish between the truth-values of what is true and what is false. I call the first, for short, the True; and the second, the False. (Frege 1891, p. 28)

On this view, therefore, in a sentence such as 'Caesar conquered Gaul', 'Caesar' is a proper name and 'conquered Gaul' is a 'concept word'. The name refers to the object for which it stands, and the 'concept word' refers to the concept for which it stands, and this concept is itself a function mapping objets onto values. Thus the concept for which '_____ conquered Gaul' stands maps the man Caesar onto that which is denoted by the sentence 'Caesar conquered Gaul'. But what is denoted by this sentence? Since the sentence is true, for Frege it denotes the truth-value True.

An Account of the First Two Parts of Frege's *Begriffsschrift*

Frege's function/argument analysis of sentences goes beyond the Aristotelian and Boolean analysis. The function

'_____ outlawed _____'

needs to be saturated at each end to turn it into a sentence: it is a function of two arguments. Frege says,

> If we imagine that in a function a symbol, which has so far been regarded as not replaceable, is now replaceable at some or all of the places where it occurs, we then obtain, by considering it in this way, a function with another argument besides the one it had before. In this way, functions of two or more arguments arise. (Frege 1879, p. 128)

It follows that, in this kind of function, the order of occurrence of the arguments does matter. I shall take an example of a sentence given by Frege for illustration:

'Cato killed Cato'.

The sentence may be analysed into function of the argument 'Cato' in more than one way. It does matter whether 'Cato' is considered as replaceable by another argument at the first occurrence or the second occurrence or at both occurrences. The four functions that may result from such an analysis are all different. In the first,

'_____ killed Cato'

'killing Cato' is the function; in the second,

'Cato killed _____'

'being killed by Cato' is the function; in the third, the relational expression

'_____ killed _____'

in which 'killing' is the function is obtained. Here the two unsaturated occurrences may be filled by two different names by using two different letters to fill the empty places: 'X killed Y'. Finally, in the fourth,

'_____ killed oneself'

'killing oneself' is the function. Here we must show that the two unsaturated occurrences are to be completed with the same name by using the same letter to complete the empty places: 'X killed X'.

Thus, what has been effected may be described as consisting of picking out four component expressions of the sentence 'Cato killed Cato', and removing one or two occurrences of each of them to form an unsaturated expression which can be viewed as a component, whose argument-places were filled by the components that have been removed.

This way at looking at the third and the fourth functions, which result from the analysis of the sentence 'Cato killed Cato', shows clearly the difficulty of the Aristotelian or Boolean analysis of proposition in terms of subject and predicate. For Frege's analysis pays no attention to the distinction between subject and predicate. Rather, what we have there is a relational expression in which the two concepts of murder and suicide can be distinguished by using letters or variables (a term which Frege dislikes) to differentiate between 'X killed Y' and 'X killed X'. Certainly, this analysis expresses more of the complexity of the sentence in question than the Aristotelian or Boolean's analysis. Regarding Frege's aim, Weiner stresses the importance of a dyadic function as follows:

> The introduction of two-place functions as constituents of conceptual contents allows Frege to express the sort of complexity that is needed if he is to be able to show that general truths about sequences can be derived using logic alone. (Weiner 1999, p. 40)

Indeed, the complicated relational inferences occurring in the mathematical notion of following in a series require dyadic functions in order to be perspicuously represented.

Although in this process of 'functional abstraction' the value of a function must be determinate for a given argument, Frege talks of 'an indeterminate function' in order to express general claims about functions. He writes,

> In order to express an indeterminate function of the argument A, we put A in parentheses following a letter, for example:
>
> $$\Phi(A).$$

Similarly,

$$\Psi(A, B)$$

> represents a function (not more explicitly determined) of the two arguments A and B. Here, the places of A and B in the parentheses represent the positions that A and B occupy in the function, regardless of whether A or B each occupies one places or more [in that function].

An Account of the First Two Parts of Frege's *Begriffsschrift* 165

Thus, in general
$$\psi(A, B) \text{ and } \psi(B, A)$$
are different. (Frege 1879, p. 129)

Accordingly, Frege describes the distinction between function and argument in a very abstract and formal way. As it has been done by logicians before him, he uses schematic letters to bring out the logical structure of propositions. In order to symbolise a function of the argument A, he employs a Greek letter Φ followed by an 'A' in parentheses: $\Phi(A)$ which signifies that A has the property Φ. A function of two arguments A and B is written $\Psi(A, B)$, where the places of A and B within the parentheses stand for the places occupied by A and B in the function which signifies that 'B stands in the Ψ-relation to A'.

In addition, Frege says that since the symbol 'Φ' occurs at a particular place in the expression '$\Phi(A)$', and since we may conceive of it as replaced by a symbol, such as 'Ψ', which then expresses another function of the argument A, we can consider $\Phi(A)$ as a function of the argument Φ (Frege 1879, p. 129). This is a shift to a different level of analysis in that Frege is saying that at the first-level, 'Φ' is a function of the argument 'A', but if we choose the judgement may also be considered as a function of the function inserted in it. Indeed, for Frege a function occurs as an argument to another function only if the expression of the second fills the gap in the incomplete expression of the first and so produces a complete propositional expression. The expression for such a function is found in Frege's way of representing generality in a proposition. Consider the proposition 'Every positive integer can be represented as the sum of four squares' which is to be split up into function and argument as follows: 'whatever arbitrary positive integer you take as argument for the function "... can be represented as the sum of four squares", the resulting proposition is always true'. Assuming that we are dealing with the universe of positive integers, the symbolical expression of the proposition is:

$$(\forall x)(\Phi(x))$$

with Φ standing for 'can be represented as the sum of four squares'. The symbol '$\forall x$' is the universal quantifier. The second 'x' is a variable indicating a place in which for the first-level function Φ an argument may be inserted. The function symbol Φ itself may also be considered as an argument; on this way of analysing it, the whole proposition may be considered as the value of the function,
$$(\forall x)(\ldots x)$$
for the argument,
$$\Phi.$$

This function having a function as its argument is what is called a second-level function.[12]

Thus, although Frege borrows his notation from functional analysis, it does not follow that he restricts himself to functions whose values are numbers. In the *Begriffsschrift*, he is concerned with linguistic functions having words as arguments and sentences as values. In his later work he takes it that a function is not just something whose values and arguments are mathematical objects, but can have values and arguments of any kind. The notion of function which is rooted in mathematical analysis is therefore extended to allow a perspicuous representation of logical relationships. Hence, Frege is quite right to say that the concept of function in Analysis, which in general he has followed, is far more restricted than the one he developed (Frege 1879, p. 129).

In the *Begriffsschrift,* whilst comparing the two propositions 'The number 20 can be represented as the sum of four squares' and 'Every positive integer can be represented as the sum of four squares', Frege warns against an illusion to which the use of ordinary language can easily give rise—that of considering the expressions 'the number 20' and 'every positive integer' as the subjects of their respective propositions. He argues:

> If we compare the two propositions:
>
> 'The number 20 can be represented as
> the sum of four squares'
>
> and
>
> 'Every positive integer can be represented as
> the sum of four squares.'
>
> it appears possible to consider 'being representable as the sum of four squares' as a function whose argument is 'the number 20' one time, and 'every positive integer' the other time. We can discern the error of this view from the observation that 'the number 20' and 'every positive integer' are not concepts of same rank. What

---

[12] In the *Begriffsschrift*, Frege does not develop the second-order predicate calculus, but in the *Grundgesetze der Arithmetik*, he devises a notation for expressing functions of the second level. This mathematical technique of logical analysis may be further pursued so as to reach a higher level of functions. Thus the abstract functional analysis contains an implicit theory of logical types. There is indeed an ascending hierarchy of types, such as first-level functions, second-level-functions, etc., and an differentiation of types at each level, such as functions of one argument, two arguments, etc.

is asserted of the number 20 cannot be asserted in the same sense of the [concept] 'every positive integer'; though, of course, in some circumstances it may be asserted of every positive integer. The expression 'every positive integer' by itself, unlike [the expression] 'the number 20', yields no independent idea; it acquires a sense only in the context of a sentence. (Frege 1879, pp. 127–128)

Unlike Boole who contends for the view that singular propositions are 'truly universals' (Boole 1847, 59f) and so follows Aristotle, Frege points out here the difference between singular propositions, involving singular terms, and general propositions. Thus Frege rejects the traditional treatment of singular propositions as universal categoricals. Moreover, Frege recognises singulars as the fundamental unit of predication and uses it to bridge the gap between Boole's primary and secondary logic (see subsection **6.1.4**).

However, Frege's claim that there is a difference between 'the number 20' and 'every positive integer' and they do not represent concepts of the same rank should be discussed. Frege observes that (1) what is asserted of 'the number 20 cannot be asserted in the same sense of the concept 'every positive integer'; and (2) the expression 'every positive integer' by itself, unlike the expression 'the number 20', yields no independent idea; it acquires a sense only in the context of a sentence. Frege may be interpreted as having in mind in (1) to introduce the difference between concepts of first and second level; and in (2) to mark as different dependent and independent entities.

This latter interpretation is problematic. Frege claims, in effect, that 'every positive integer' acquires a sense only in the context of a sentence because it is a dependent expression. But does not the same apply to 'the number 20'? It is striking that Frege considers here 'the number 20' as an independent idea which does not acquire a meaning only from the context of the sentence. For, in *The Foundations of Arithmetic*, he states as a principle 'never try to define the meaning of a word in isolation, but as it is used in the context of a proposition'. And subsequently he contends that 'only by adhering to this can we avoid a physical view of number without slipping into a psychological view of it' (Frege 1884, p. 116). The concept of number is then obtained by determining the sense of a proposition in which a number word occurs, or by fixing the sense of a numerical identity.

Frege may be interpreted in this way: 'the number 20' has an independent denotation (reference), unlike 'every positive integer'. How to show this? By pointing out that 'Every positive integer is even or odd' is not equivalent to 'Every positive integer is even' or 'Every positive integer is odd'. But 'The number 20 is even or odd' is equivalent to 'The number 20 is even' or 'the number 20 is odd'. In effect, Frege is paving the way for his analysis of universal propositions. It is his theory of quantification which explains how such propositions should be understood. Although the expression 'every positive integer' stands for a concept, it stands for something that needs further analysis. The logical structure of the concept 'every positive integer' (i.e. for all x, if x is a positive integer) is concealed by ordinary language.

If the above interpretation is correct, then it may be said that when Frege made his observations, he thought mainly of the notion of generality. And since his analysis of this notion involves the universal quantifier, his observations may be taken as an anticipation of his later distinction between first and second level concepts.

5.3.7 Frege's Theory of Quantification

The theory of quantification stems from Frege's theory of functions. It constitutes the technical apparatus with which Frege expresses perspicuously the generality involved in mathematical propositions such as, 'Every even number is the sum of two primes'. Thus, he provides mathematicians with a method of proof which allows them to shun intuition in their inferences, and to represent them in a more precise way than can be done with ordinary language and mathematical symbols. Frege then builds up a new logic capable of expressing judgements with multiple generality occurring in mathematical proofs.

Aristotelian logic was concerned with syllogisms, and set up a number of rules in order to work out a valid inference. However, syllogistic theory was unable to provide an adequate representation of all inferences containing words such as, 'every' or 'any' (or 'all' and 'some'). For the focus of reasoning was on inferences about types of thing. For instance, the proposition 'All men are mortal' was about men and mortal things. Frege, by contrast, considers as the fundamental form of a general proposition a proposition about everything. Thus, in his account, the proposition 'All men are mortal' is not about men at all, but one about everything—to the effect that anything is such that 'if it is a man, then it is mortal'.

It follows that a general proposition such as 'All men are mortal' is a complex proposition which can only be analysed by means of the quantifier notation and its associated variable. Furthermore, when it is understood that 'All men are mortal' is equivalent to the universal closure of the functional expression 'if x is a man then x is mortal', every proposition representable in Aristotelian syllogistic can be represented perspicuously. This is achievable because Frege reverses the order of priority Boole ascribes to his 'primary' and 'secondary' propositions. The crucial point is that Frege regards judgements as prior to concepts (see subsection **6.1.4**).

An Account of the First Two Parts of Frege's *Begriffsschrift*

The device of quantification is carried out in sections 11 and 12 of *Begriffsschrift*, where Frege sets up a way in which the complexity of a proposition is accounted for in terms of variables and quantifiers. He writes:

> In the expression of a judgement, we can always regard the combination of symbols to the right of
>
> ├────────
>
> as a function of one of the symbols occurring in it. *If we replace this argument by a German letter and introduce in the content stroke a concavity containing the same German letter, as in*
>
>  $\Phi(\mathfrak{a})$
>
> *then this stands for the judgement that the function is a fact whatever we may take as its argument.* (Frege 1979, p. 130)

In this formula, the German or gothic letters stand for the variables and the concavity which occurs within the content stroke expresses: 'whatever we may take'. This 'concavity' is now known as 'the universal quantifier'. The above formula is saying that every judgement of the form $\Phi(x)$ is to be affirmed, and may be written in the modern language of predicate calculus as follows:

$$(\forall x)\Phi(x)$$

I shall consider again the following general proposition in order to throw more light on Frege's notation for generality:

'Every positive integer can be represented as the sum of four squares.'

It may be expressed as follows:

'For every $\mathfrak{a}$, if $\mathfrak{a}$ is a positive integer,
then $\mathfrak{a}$ is representable as the sum of four squares.'

The quantifier expression begins the expression here. The second and the third German or gothic letter are places that could be filled by a positive integer name. But the German or gothic letter is a variable, not a positive integer name. Hence, the German or gothic letter does not here stand for some particular object, instead it expresses generality when put together with a quantifier. It is a bound variable since its occurrences in the expression are all tied to the quantifier expression that begins the expression. In other words, the quantifier expression has the entire expression in its scope.[13] In Frege's notation our general proposition is represented as

$$\vdash \underset{\mathfrak{a}}{\bigcap} \begin{array}{l} \Phi(\mathfrak{a}) \\ \Psi(\mathfrak{a}) \end{array}$$

with Ψ and Φ standing for respectively the functions, 'being a positive integer' and 'being representable as the sum of four squares'. In modern notation, its symbolisation is:

$(\forall x)(x$ is a positive integer $\Rightarrow x$ is representable as the sum of four squares),

or,

$$(\forall x)(\Psi(x) \Rightarrow \Phi(x)).$$

This way of expressing generality points out two important features, which are the use of variables and of quantifiers which delimit the scope of the generality. What Frege does is to introduce a new symbol, that is, the 'concavity' over which stands a variable, so as to indicate scope. According to him the concavity with the German or gothic letter written in it is necessary, in so far as *'it delimits the scope of the generality signified by the letter. The German letter retains its significance only within its scope'* (Frege 1879, p. 131). Moreover, there is what is called 'multiple generality', and the symbols for generality themselves may have different scope. As Frege says it, 'the scope of one German letter can include that of another' (Frege 1879, p. 131). This occurs frequently in the third part of the *Begriffsschrift* in which Frege deals with the mathematical notion of sequences involving multiple generality. The variables and the quantifiers allow Frege's logic to handle polyadic predicates and nested quantification, and thus make it more powerful than all the great logical systems of the past.

---

[13] A quantifier expression that has the entire expression in its scope is particularly important in that it allows substitution. Thus from the sentence 'If everything is wet we cannot light a fire' (Kneale 1962, p. 487), which, when written in Frege's notation, would include a universal quantifier that does not have the entire sentence in its scope, one cannot derive, by substitution, 'If the grass is wet we cannot light a fire'.(maybe the fire can be lighted, so long as the wood is dry). But from the proposition 'All men are mortal', which, when written in Frege's notation include a universal quantifier that has the entire proposition in its scope, one can derive by substitution 'If Socrates is a man then Socrates is mortal'. Frege then introduces in his notation italic letters to designate letters not linked to any explicit quantifier. In the terminology of modern logicians, italic letters are called free variables (see subsection **5.4.2**).

In effect, as mentioned in the previous subsection, the most important weakness of logic before Frege was its inability to handle sentences involving expressions for multiple generality which are abundant in mathematics and are often ambiguous. The theory of 'supposition', which is developed by Medieval logicians, was intended, in part to solve problems involving multiple generality i.e. how to distinguish 'Every donkey of some man is running' from 'Some donkey of every man is running'[14]. The theory considered a term i.e. 'men' as referring to individual men, such as Socrates and Plato, but referring to them in different ways determined by the different structures of the sentences in which it occurs. General terms i.e. 'men' and 'mortal' referred to members of classes. As Boehner says, 'supposition deals for the most part with the extension or range of predicates in reference to individuals' (Boehner 1952, p. 28).

But, although the theory of supposition had to take into account semantical considerations, it did not use artificial language and was satisfied with the clarification and determination of the structures of the Latin language. It could not therefore provide a satisfactory account of sentences containing expressions for multiple generality[15]. For, as Dummett points out, the difficulty of the medieval theory of *suppositio* arose out of trying to consider a sentence like 'Everybody envies somebody' as being constructed simultaneously out of its three components, the relational expression represented by the verb, and the two signs of generality (Dummett 1981b, p. 10). As a result, it fails to represent differences in the scope of quantifiers when they occur in a single sentence.

Frege, by contrast, does not account for quantifier expressions by referring to the quantity of terms. Rather, his 'proper name' is understood to refer to some given object of the appropriate domain. Frege considers a sentence involving multiple quantification as having different constructional histories, corresponding to the different symbols of generality occurring in it. This conception leads him to semantical considerations in that it requires the account of the truth-conditions of sentences containing expressions of generality. Since he aims to explain the mathematical notion of following in a series in purely logical terms, he builds up a logical language which can handle polyadic predicates and nested quantifiers. The power of this artificial language can be

---

[14] In his 1978 paper, 'Multiple Quantification and the Use of Special Quantifiers in Early Sixteenth Century Logic', Ashworth investigates the analysis which logicians in the medieval tradition gave of such sentences as 'There is somebody all of whose donkeys are running', 'Everybody has at least one donkey which is running', and 'At least one of the donkeys which everybody owns is running'. His discussion is based on the work of a group of logicians who were at the University of Paris in the first two decades of the sixteenth century, in particular Fernando de Enzinas, Antonio Coronel, and Domingo de Soto.

[15] Currie points out the weakness of the theory of *suppositio* (see Currie 1982, pp. 25–26). See also the accounts in Boehner (1952), Moody (1953), Geach (1962) and Ashworth (1978).

evaluated by looking at an example of dyadic quantifiers in which two occurrences of quantifying expressions are connected by a relational expression. Let us consider the multiply general sentence:

'Everybody loves somebody'.

The sentence is ambiguous in that it may be interpreted in two different ways: it may be understood as saying that for everybody there is someone to love, or as saying somebody is loved by everybody. The elucidation of the ambiguity of the sentence requires the understanding of its two different constructional histories. In what follows, the truth-condition of the sentence in its interpretation corresponding to 'for everybody there is someone to love' will be accounted for in a Fregean way by first interpreting the universal quantifier and then dealing with the existential quantifier. Thus the sentence would be viewed as being constructed by first inserting the existential quantifier and then the universal quantifier. So the order of interpretation reverses the order of construction. It would be represented in the quantifier notation in such a way that the scope of the universal quantifier would include that of the existential quantifier.

In Frege's logical language a sentence may be constructed by putting together a quantifier and a monadic predicate. Then the function / argument analysis allows one to consider the monadic predicate itself as having been constructed from a sentence by removing one or more occurrences of some one singular term i.e. proper name. Thus suppose we begin with a sentence such as

'Abraham loves Sara'

and remove 'Sara' to yield the functional expression

'Abraham loves _____'.

We can then insert the sign of generality 'somebody' into the blank, yielding

'Abraham loves Somebody'.

This resulting sentence may be, in turn, subjected to the same process by removing 'Abraham' from the sentence and inserting the sign of generality 'Everybody', and thus arriving at our original sentence

'Everybody loves Somebody'.

In order to describe syntactically the sentence, it is necessary to introduce a symbolic notation for existential quantification i.e. $(\exists x)$ in modern notation, which stands for the expression of generality 'somebody' as the universal quantifier does for the expression of generality 'everybody'. The contribution of this existential quantifier to the truth-value of an entire sentence may be given by a rule that states the conditions under which the corresponding functional expression, suitably completed, must produce truth (see Frege 1893, pp. 42–43). This rule states that

$$(\exists x)\Phi(x)$$

stands for the judgement that the function $\phi(x)$ is a fact for some argument[16]. We already know that a sentence containing the universal quantifier is true if and only if the functional expression is a fact whatever we take as its argument. Regarding a monadic predicate, it is true of a given individual if and only if the sentence which results from inserting a name of that individual in the gap of the predicate is true.

From these simple rules we can now provide a Fregean account of the truth-conditions of the sentence 'Everybody loves Somebody' corresponding to its above constructional history by dealing with the expressions of generality in the inverse order of their occurrence in the sentence. Thus, 'Everybody loves Somebody' is true if and only if each of the sentences, 'Abraham loves somebody', 'Ismhael loves somebody', 'Isaac loves somebody' and so on, is true; and 'Abraham loves somebody' is, in turn, true if and only if at least one of the sentences, 'Abraham loves Sara', 'Abraham loves Hagar', 'Abraham loves Keturah' and so on, is true. As this example shows, there is an ad hoc convention which we tacitly employ, that is, 'the order of construction corresponds to the inverse order of occurrence of the signs of generality in the sentence: when "everybody" precedes "somebody", it is taken as having been introduced later in the step-by-step construction, and conversely' (Dummett 1981b, p. 12, see also pp. 8–33, to which this treatment of Frege's account of generality is indebted, for useful elaboration).

It is, of course, also possible to interpret 'Everybody loves Somebody' in another way, as in effect 'Somebody is loved by everybody' (writing it in accordance with Dummett ad hoc convention). On this interpretation, the order of construction is reversed ('everybody' is inserted before 'somebody') and as a result, the order of interpretation of the quantifiers is likewise reversed.

What is comprehended from the above expression of the power of Frege's device of quantification can be recapitulated in three points: first Frege's account of generality shows how the truth-conditions of general sentences depend upon those of the corresponding singular sentences subjected to the function

---

[16]In section 12 of the *Begriffsschrift* as well as in section 8 of the *Grundgesetze der Arithmetik*, Frege combines the sign of negation and the universal quantifier so as to represent the existential quantifier. He also accounts for its truth-condition.

/ argument analysis; secondly this account proceeds step-by-step and as the construction of the sentence goes on, the expression of generality decreases at each stage; and finally, the application of Frege's account to natural language requires the consideration of the sentence as having a constructional history, that is, as being constructed in stages.

It is now possible to understand the notion of the 'scope' of a quantifier. The two interpretations of the sentence 'Everybody loves Somebody' involve two different scopes of the quantifiers. In the interpretation matching the constructional history in which the universal quantifier is inserted last (For everybody there is someone to love), the existential quantifier lies within the scope of the universal quantifier, whereas in the interpretation matching the constructional history in which the existential quantifier is inserted last (Somebody is loved by everybody), the universal quantifier lies within the scope of the existential quantifier. When a quantifier lies within the scope of another, there results a kind of complexity that was never spelled out by the great logicians of the past. As Dudman puts it,

> The essence of Frege's advance beyond Boole was the appearance, in 1879, of truthfunctional connectives within the scope of quantifiers, along with quantifiers within the scope of truthfunctional connectives. But simultaneously there occurred a yet greater advance: the appearance of quantifiers within the scope of quantifiers (Dudman 1976, p. 137).

Indeed, the appearance of quantifiers within the scope of quantifiers allows Frege to represent perspicuously relations of considerable complexity as the complicated inferences in mathematics and scientific discourse whatsoever. For example, with '$Pr(y)$' standing for 'y is a prime number' and 'x' for a number, the well-known mathematical judgement, 'there is a prime number greater than any given number', can be represented in modern notation as follows:

$$(\forall x)(\exists y)(Pr(y) \land (y > x)).$$

Certainly, the handling of the nested quantifiers $(\forall x)(\exists y)$ goes beyond anything to be found in Boole. Frege also employs the notion of the scope of the quantifier in Part III of the *Begriffsschrift* in order to define 'y follows x in the f-sequence' (see Frege 1879, formula 69, section 24, p. 167). This definition plays an important rôle in Frege's aim to reduce mathematical induction to logical inference.

So far what has been done is to apply Frege's account of generality to ordinary language sentences and to show how it can elucidate the ambiguities which teem in them. But, by replacing the notation used to express generality in ordinary language with a new notation—that of quantifiers and variables, Frege makes impossible the occurrence of such ambiguities in his logical language. As Dummett says, 'the point of this new notation was to enable the constructional history of any sentence to be determined unambiguously' (Dummett 1981b, p. 12). Indeed, Frege's new notation which assigns a different logical form to each interpretation of the sentence 'Everybody loves Somebody' clears up the ambiguity. In Frege's notation the interpretation corresponding to 'For everybody there is someone to love', appears as

$$\neg\underset{\mathfrak{a}}{\smile}\neg\underset{\mathfrak{b}}{\smile}\neg L(\mathfrak{a}, \mathfrak{b})$$

Whilst the other interpretation corresponding to 'Somebody is loved by everybody' is written

$$\neg\underset{\mathfrak{b}}{\smile}\neg\underset{\mathfrak{a}}{\smile}L(\mathfrak{a}, \mathfrak{b})$$

The first formula will be represented in the modern equivalent of Frege's notation as

$$(\forall x)(\exists y)Lxy$$

in which the different variables ('x' and 'y') indicate clearly the scopes of the two quantifiers. The second formula will be written as

$$(\exists x)(\forall y)Lxy$$

It should be noted that while dealing with the step-by step construction of the sentence 'Everybody loves Somebody' a symbol for existential quantification was introduced. However, Frege himself does not introduce a special symbol to represent the word 'some'. For it does not seem indispensable to have a special notation for the existential quantifier, in so far as it had long been assumed by some logicians that a sentence such as 'some swans are black' is equivalent to 'not all swans are not black.' In other words, saying that something has a particular property is denying that everything fails to have the same property. Frege makes use of this relationship between 'some' and 'not all...not' in order to express an 'existential' proposition. Thus, 'some swans are black' may be represented in Frege's notation by means of the universal quantifier and the negation sign as follows:

$$\vdash\!\!-\!\!\overset{\mathfrak{a}}{\smile}\!\!\top\!\!-\varphi(\mathfrak{a})$$

With φ standing for the function, 'being a black swan', the formula means: there are black swans or is at least one black swan. In modern notation the formula is equivalent to

$$\neg(\forall x)\neg\varphi(x)$$

Accordingly, Frege is not only capable of delimiting the scope of general contents; he is also capable of expressing particular as well as existential judgements by combining his 'quantifier' with the negation sign. I shall show how Frege applies such a combination to propositions of restricted generality such as the type of proposition in Aristotelian logic.

Indeed, Frege shows his ability to account for restricted generality in terms of unrestricted generality and truth-functions with the four types of proposition found in Aristotelian square of opposition (Frege 1879, pp. 133-135). Thus, he notes that 'Every X is a P', or 'All X's are P's' is equivalent to 'If something has the property X, then it also has the property P'. And this is expressed in Frege's notation as follows:

$$\vdash\!\!-\!\!\overset{\mathfrak{a}}{\smile}\!\!\top\!\!\begin{array}{l} P(\mathfrak{a}) \\ X(\mathfrak{a}) \end{array}$$

The formula will be true if nothing has the property X, i.e. if there are no Xs.

'No Ψ is a P' is equivalent to 'What has the property Ψ does not have the property P'. It is expressed as:

$$\vdash\!\!-\!\!\overset{\mathfrak{a}}{\smile}\!\!\top\!\!\begin{array}{l} P(\mathfrak{a}) \\ \Psi(\mathfrak{a}) \end{array}$$

'Some Λs are not Ps' is equivalent to 'It is not the case that everything which is Λ is a P'. Its notation appears as:

$$\vdash\!\!-\!\!\overset{\mathfrak{a}}{\smile}\!\!\top\!\!\begin{array}{l} P(\mathfrak{a}) \\ \Lambda(\mathfrak{a}) \end{array}$$

Finally, 'Some Ms are Ps', which is equivalent to 'It is possible for a M to be a P' is represented as:

$$\vdash\!\!-\!\!\overset{\mathfrak{a}}{\smile}\!\!\top\!\!\begin{array}{l} P(\mathfrak{a}) \\ M(\mathfrak{a}) \end{array}$$

An Account of the First Two Parts of Frege's *Begriffsschrift* 177

Frege (1879, p. 135) then tabulates the square of logical opposition as:

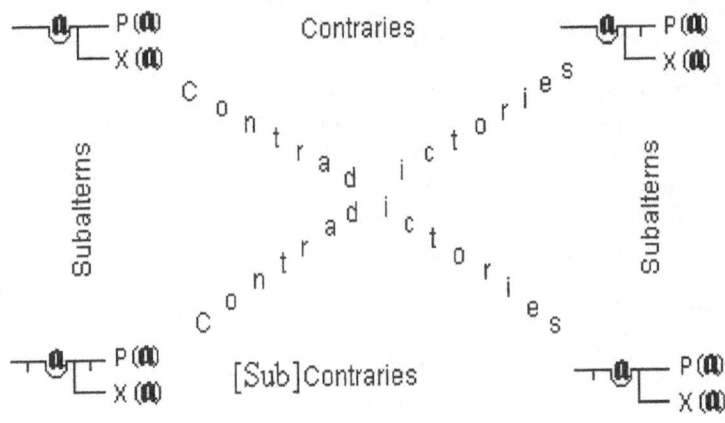

Figure 5.1

These four types of Aristotelian propositions named **A**, **E**, **I**, **O** can be expressed in modern notation as follows:

A: $(\forall x)(X(x) \Rightarrow P(x))$

E: $(\forall x)(X(x) \Rightarrow \neg P(x))$

I: $\neg(\forall x)(X(x) \Rightarrow \neg P(x))$ or $(\exists x)(X(x) \, \& \, P(x))$

O: $\neg(\forall x)(X(x) \Rightarrow P(x))$ or $(\exists x)(X(x) \, \& \, \neg P(x))$.

What is striking here, and then should be stressed, is the fact that Frege curiously pictures the relations between the four propositions as they stand in Aristotelian logic without noticing that contrariety, subcontrariety and subordination are all invalid immediate inferences within his formal proof. In the Aristotelian representation of the square of logical opposition **A** and **E** are contraries, which means that they cannot both be true but may be both false. **I** and **O** are subcontraries, that is, they cannot both be false but may both be true. **A** is the subaltern of **I** which means that **I** cannot be true without **A** being true, but **A** may be true without **I** being true.

What Frege fails to note is that in his system **A** and **E** can both be true, if there are no Xs; **I** and **O** can both be false, if there are no Xs; and **I** does not follow from **A**, and **O** does not follow from **E**. For example, according to the rule of Aristotelian logic 'All X is P' and 'No X is P' cannot both be true. The expressions of these in Frege's system are respectively:

$$(\forall x)(X(x) \Rightarrow P(x))$$
$$(\forall x)(X(x) \Rightarrow \neg P(x))$$

But if there are no Xs as in the proposition 'All Sirens are seductive', then by the truth-table definition of the implication both these propositions are true, or to use the current expression are 'vacuously true.' Hence there is a case in which contraries are both true. The rule of subcontrariety states that 'Some X is P' and 'Some X is not P' may both be true, but cannot both be false. However, if there are no Xs, then both subcontraries are false. The rule of subordinates is that, if 'All X is P'is true, then so is 'Some$X$is P'. But if there no $X's$, then 'All X is P' and 'No$X$is$P'$are 'vacuously true', while 'Some$X$ is P' and 'Some X is not$P$' are false. Accordingly, the only rule of the square of logical opposition which remains valid in Frege's logical system is the law of contradictories (see subsection **4.1.1**).

In the footnote 21 of his later work, *Grundgesetze der Arithmetik*, Frege notes that 'many logicians seem to assume without ado that concepts are realized, and to overlook entirely the very important case of empty concepts, perhaps because they quite wrongly do not recognize empty concepts as justified' (Frege 1893, p. 56). But, in the *Begriffsschrift*, when Frege concerned with the type of proposition dealt with in Aristotelian logic, unfortunately he reproduced the square of opposition as it stands and *ipso facto* gave the impression of being amongst these logicians who 'wrongly did not recognize empty concepts as justified'.

On this issue, it is worth noting that Frege and Boole had a similar attitude. Both of them were committed in their logical systems to declining the existential import of universals propositions, but neither of them realised when writing that this was a significant point that needed to be highlighted.

5.4 A Formal System of Logic

After having built up the formula language as a tool specially devised for mathematicians to render their definitions and inferences in greater exactitude, Frege sets up logic as an axiomatic science, in which inferences that have been accepted as valid by means of Aristotelian or propositional logic may also be proved as valid. He develops a formal system of logic which proceeds by primitive symbols, rules of combination of these symbols, axioms, and rules of inference proving the theorems. In this way, Frege follows the Euclidean procedure of setting up geometry as an axiomatic science.

I shall first discuss the epistemological issue underlying the quest of rigour and precision in science, that is, the investigation and the identification of the ultimate foundations of indemonstrable truths. Then, I shall present Frege's axiomatisation of propositional calculus.

5.4.1 A Euclidean Procedure

In the first section of *The Foundations of Arithmetic*, Frege writes:

> After deserting for a time the old Euclidean standards of rigour, mathematics is now returning to them, and even making efforts to go beyond them... Proof is now demanded of many things that formerly passed as self-evident. Again and again the limits of validity of a proposition have been in this way established for the first time. ... In all directions these same ideals can be seen at work—rigour of proof, precise delimitation of extent of validity, and as a means to this, sharp definition of concepts. (Frege 1884, p. 1)

By advocating a Euclidean procedure, Frege underlines an epistemological issue: the status of logical axioms. Frege holds that true knowledge such as logic must imitate Euclidean procedure. There is an exact analogy between logic and geometry regarding the nature and the way in which the primitive logical axioms are carried out. The axioms of geometry are a generalisation expressing a relation between concepts whose theorems are a demonstration of the kernel of its primitive concepts. Likewise, the axioms and the theorems of Frege's propositional calculus elucidate the nature of the primitive logical constants such as the indefinable symbols of negation and conditionality.

But like geometrical axioms, the logical axioms are not drawn from definitions of concepts, since Frege conceives them as indemonstrable truths. They cannot be drawn from a definition of 'truth' either, because, in his eyes, the notion of 'truth' itself is indefinable. Instead, logic, as he sees it, is a process of justifying truths: it justifies the truth of propositions from propositions whose truths are clearly recognised. What is distinctive about logic (as compared with geometry) is its absolute universality.

In an essay *On Euclidean Geometry*, Frege writes:

> No thought that is held to be false can be accepted as an axiom, for an axiom is a truth. Furthermore, it is part of the concept of an axiom that it can be recognized as true independently of other truth. (Truths can be inferred in accordance with the logical laws of inference.) If a truth is given, it can be asked from what other truths its truth follows in accordance with the logical laws

of inference. When this question has been answered, we can go on to ask of each of the truths that we have thus discovered from what other truths its truth follows in accordance with the laws of inference. (Frege 1899/1906, p. 168)

Thus, logical justifications may be pursued until axioms are reached. That is why Frege says that 'there must be judgements whose justification rests on something else, if they stand in need of justification at all' (Frege 1879/91, p. 3). There epistemology comes in, so as to account for the 'something else' upon which rest some judgements that cannot be justified by inferring them from other judgements.

In the preface of *Begriffsschrift*, Frege distinguishes between two ways of grasping knowledge, an inquiry into the 'context of discovery' of a particular form of knowledge, and an inquiry concerning its justification. He claims:

The apprehension of a scientific truth proceeds, as a rule, through several stages of certainty. First guessed, perhaps, from an inadequate number of particular cases, a universal proposition becomes little by little more firmly established by obtaining, through chains of reasoning, a connection with other truths - whether conclusions which find confirmation in some other way are derived from it; or, conversely, whether it comes to be seen as a conclusion from already established propositions. (Frege 1879, p. 103)

There follow two questions: on the one hand, it may be asked in what way a proposition is progressively obtained, and on the other hand, in what way it is to be most securely established. It is the task of a theory of knowledge to answer to the second question, and Frege holds that 'its answer is connected with the inner nature of the proposition concerned'. A proposition can be justified whilst recognising that there is a limit beyond which we have to stop establishing it as true. Otherwise, we would be lost in a *regressus in infinitum*. As he puts it: ' the question why and with what right we acknowledge a law of logic to be true, logic can answer only by reducing it to another law of logic. Where that is not possible, logic can give no answer' (Frege 1893, I, XVI). Where that is not possible we arrive at the primitive logical axioms, which then must be accepted without a purely logical justification.

Although Frege recognises that epistemology aims to identify the ultimate grounds of these primitive logical axioms, it was not actually his concern to illuminate the issue. Hence, Weiner is correct when saying that 'the point at which epistemology, on Frege's view, comes in to an investigation of justification is precisely the point at which, in his own discussions of justification, Frege is mute.' He admittedly claims that 'it would not perhaps be beside the mark to say that the laws of logic are nothing other than an unfolding of the content of the word 'true'. Anyone who has failed to grasp the meaning of this word-what marks it off from others-cannot attain to any clear idea of what the task of logic is' (Frege 1879/91, p. 3). But we cannot take from this statement any clear method for identifying the 'basic' laws of logic. Hence like the Euclidean axioms, the grounds which justify the recognition of a truth express an indemonstrable truth that has to be immediately recognised.

However, although Frege appeals to the Euclidean model, he goes beyond it so as to perfect it. For,

> Even such a conscientious and rigorous writer as Euclid often makes tacit use of presuppositions which he specifies neither in his axioms and postulates nor in the premisses of the particular theorem.... Only by paying particular attention, however, can the reader become aware of the omission of these sentences, especially since they seem so close to being as fundamental as the laws of thought that they are used just like those laws themselves. (Frege 1882b, p. 85)

This is tantamount to saying that if the Euclidean proofs were given in Fregean formula language, then the omission of these sentences would be apparent. For the inference carried out by his logical language may be seen as a logical consequence of the laws of thought. But in the Euclidean proofs there seems to be no need for following rules of inference. Moreover, the axioms of logic are justified immediately by a 'logical source of knowledge which is wholly inside us and thus appears to be more proof against contamination' due to the mistakes that may stem from the senses, which yet provide the axioms of geometry (Frege 1924/25b, p. 269).

Consequently, in order to gain more rigour and precision, Frege requires logic to be presented as an axiomatic deductive system, in which the primitive propositions are explicit together with the rules which govern and regulate the chain of inference proving the theorems. The Fregean formula language does not let through anything that was not expressly presupposed, even the propositions that 'seem so close to being as fundamental as the laws of thought' which we may consider so obvious that we do not need to express them explicitly.

5.4.2 The Axiomatisation of Propositional Calculus

The requirement of rigour in the axiomatisation of logical propositions is exhibited in the chapter of *Begriffsschrift* entitled 'Representation and Derivation of Some Judgments of Pure Thought'. There Frege sets up the systematic formulation of propositional calculus by presenting it in an axiomatic form, from which all laws of logic are inferred in accordance with only two primitive signs (negation and conditional), one rule of inference (*modus ponens* and an implicit rule of substitution) and nine primitive axioms. In fact, Frege also employs a further rule of inference: an implicit rule of substitution.

The Axioms

Three of the axioms require only the variable letters and the sign for conditionality (the formulae **1**, **2**, and **8**); another three require the sign of negation (formulae **28**, **31** and **41**); the last three axioms require the sign for identity of content and generality. The formulae are exhibited in *Begriffsschrift* as follows:

The modern notation of these expressions may be given as follows:

(1) $\vdash a \Rightarrow (b \Rightarrow a)$,

(2) $\vdash (c \Rightarrow (b \Rightarrow a)) \Rightarrow ((c \Rightarrow b) \Rightarrow (c \Rightarrow a))$,

(8) $\vdash (d \Rightarrow (b \Rightarrow a)) \Rightarrow (b \Rightarrow (d \Rightarrow a))$,

(28) $\vdash (b \Rightarrow a) \Rightarrow (\neg a \Rightarrow \neg b)$,

(31) $\vdash \neg\neg a \Rightarrow a$,

(41) $\vdash a \Rightarrow \neg\neg a$,

(52) $\vdash (c = d) \Rightarrow (f(c) \Rightarrow f(d))$,

(**54**) ⊢ $c = c$,

(**58**) ⊢ $(\forall \mathfrak{a} f(\mathfrak{a})) \Rightarrow f(c)$

Frege employs italic letters to express all the axioms. For him 'an italic letter is always to have as its scope the content of the whole judgement, and this need not be signified by a concavity in the content stroke...' (Frege 1879, p. 131). In the terminology of modern logicians, italic letters are called free variables. An occurrence of a variable in a proposition is called free in that proposition when it neither occurs in, nor is it bound by, any quantifier within the limits of the given proposition. Thus the occurrences of the variables, 'a', 'b' 'c' and 'd' in the axioms are free therein, but the occurrences of the gothic letter in the ninth axiom are bound, that is, not free, in this propositional function. These variables whose ranges are the truth-values, and for which sentences expressing propositions may be substituted are called propositional variables. In the modern notation of the three last axioms, we take the letter 'f' as an abbreviation for predicate expressions; and the variables 'c', 'd' as individual variables. But it is worth noting that in his use of italic letters Frege does not distinguish between propositional variables and individual variables.

The first six axioms may be regarded as appertaining to the calculus of propositions, which is an independent theory, and the most fundamental part of logic upon which the calculus of propositional functions depends, so as to be formulated systematically. The ninth, in which the quantifier appears, is required for the logic of quantifiers. The seventh and eighth axioms appertain to the theory of identity.

Rules of Inference

There is a difference between the axioms and the rules of inference within an axiomatic system in that an axiom constitutes a logical truth of the system, whereas a rule of inference is not itself true or false. It just allows us to derive a consequence from a proof when we have already set up some self-evident axioms. Thus, in order to infer the theorems from the axioms of his formal system of logic, Frege relies upon two rules, namely, those of detachment and substitution. The former is called *modus ponens* and is the only rule of inference he explicitly acknowledges.

Indeed, Frege utilises *modus ponens* as a rule of inference to carry out the theorems of his logical axiomatic system, and thereby to prove all the laws of logic. For him, *modus ponens* is a formal procedure, which allows a proposition to be written down, given that other propositions have been written down. The proof is purely formal, since it is the form rather than the content, which allows one to move from one proposition to another. Here is how he expresses the rule: If we have two propositions

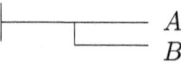

and

then a rule of inference allows to draw as a conclusion the proposition

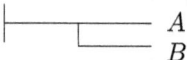

The cogency of this rule of inference requires that the same proposition be expressed by the antecedent of the conditional and by the second premise. That is why *modus ponens* is called by modern logicians 'affirming the antecedent'.

Although Frege claims that he uses only *modus ponens* as a rule of inference, he actually employs a rule of substitution, of which he shows the procedure in section 6 of *Begriffsschrift*. Indeed, it would be arduous if the proofs would always have to be written in full. Frege then figures out a way to abbreviate the procedure. It is described as follows: 'each judgement which occurs in the context of a proof is labelled with a number which is placed to the right of the judgement at its first occurrence' (Frege 1879, p. 118). As an example, he says: suppose that the judgement

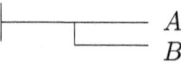

has been labelled with X. Then, Frege writes the *modus ponens* inference this way:

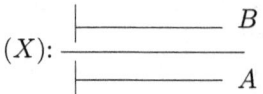

An Account of the First Two Parts of Frege's *Begriffsschrift* 185

In addition, Kneale points out the existence of two others rules of inference those of generalisation and confinement. Regarding the rule of generalisation, it follows from Frege's statement that 'an italic letter may always be replaced by a German letter which does not yet occur in the judgement; when this is done, the concavity must be placed immediately after the judgement stroke' (Frege 1879, p. 132). Thus, as Kneale notes it, the statement amounts in effect to a third rule of inference (Kneale 1962, p. 489). As an example, instead of

$$\vdash\!\!\!-\!\!\!-\!\!\!-\!\!\!-\!\!\!-\!\!\!- X(a)$$

Frege puts

$$\vdash\!\!\!-\!\!\!\cup_{\mathfrak{a}}\!\!-\!\!\!-\!\!\!- X(\mathfrak{a})$$

if a occurs only in the argument place of $X(a)$.

Frege provides also an introduction rule for the universal quantifier. Namely that if the conditional

$$\vdash\!\!\!-\!\!\!-\!\!\!\top\!\!-\!\!\! \Phi(a) \\ \qquad\qquad\;\; A$$

is affirmed, then the generalised conditional

$$\vdash\!\!\!-\!\!\!\cup_{\mathfrak{a}}\!\!\top\!\!-\!\!\! \Phi(\mathfrak{a}) \\ \qquad\qquad\qquad A$$

is also affirmed, 'if A is an expression in which a does not occur and a stands only in argument places of $\Phi(a)$' (Frege 1879, p. 132). Kneale identifies this inference as Frege's fourth rule, which he calls confinement (Kneale 1962, p. 489).

Subsequently, Kneale says that 'taken with these four rules his axioms are indeed a complete set in the technical sense of that phrase, though their sufficiency could not be demonstrated in Frege's day' (Kneale 1962, p. 489). Indeed, the axioms and rules constitute a complete set of axioms and rules for first-order predicate logic. Although in the *Begriffsschrift* there is no metamathematical investigations into the completeness, consistency, or independence of the axioms, Frege's axiomatic system addresses the issue. Already, in *Grundgesetze der Arithmetik*, he considers completeness to be indispensable for the rigour of the conduct of proof in that it does not permit the tacit attachment of presupposition in thought (Frege 1893, VI, p. 2). However, it was left to Frege's successors to show the completeness of first-order logic. According to Kneale, in *On the History of the Logic of Propositions*, Lukasiewicz later proved the completeness of Frege's axiomatisation in which every valid logical truth of two-valued propositional calculus can be demonstrated using his six axioms and the rules of inference. But the first proof of the completeness of the first-order predicate calculus was given by Gödel in his 1930 doctoral thesis.

After Frege's axiomatisation of propositional calculus, the earliest attempt to complete his system was made by Russell and Whitehead in *Principia Mathematica*, which aimed at proving that pure mathematics is nothing other than an extension of formal logic. Their work then gave an impetus to all further research in mathematical logic. Thus Hilbert and Ackermann's metamathematical considerations revolved around it. As well, Gödel's most famous paper is entitled: 'On Formally Undecidable Propositions in Principia Mathematica and Related Systems I.' This shows the great influence of Frege's research programme to which *Principia Mathematica* belongs.

5.5 The Place of *Begriffsschrift* in the History of Logic

Frege is widely considered as the greatest logician since Aristotle. In the *Begriffsschrift*, he developed the first modern system of formal logic, showing that mathematical induction may be expressed in such a way that it may be regarded as a particular case of the familiar logical forms of inferences, and thereby reducible to the laws of logic. Certainly, this is a tremendous achievement, which puts him in the logical pantheon beside Aristotle. However, I shall address a historical issue related to the extent to which *Begriffsschrift* may be considered as a novelty in the history of logic.

Undoubtedly, 1879 is a revolutionary epoch in logic. But *Begriffsschrift* may not have been so new as some readers of Frege have claimed. It represents the harvest of the work in mathematics and logic which was carried out in the second half of the nineteenth century, and which sowed seeds in the ground for the growth of mathematical logic.

Dummett, the doyen of Frege's studies, is in complete disagreement with this view. As he claims it,

> Frege's ideas appear to have no ancestry. He applied himself to formal logic, and invented a totally new approach.... It is true that his works are full of diatribes against the mistakes of others: but he never seems to have learned from anybody else, not even by reaction; other authors appear in his writings only as object-lessons in how not to handle the subject. And so, perhaps, it is vain to

wish that he had paid more attention to the work of his successors in logic and the foundations of mathematics: perhaps he was incapable of sailing any sea on which other ships were in sight. (Dummett 1981b, p. 661)

As Pallas sprang from Zeus's forehead, Frege's logic then arose *ex nihilo*! Frege would not line up with Dummett's account, in so far as he held that

> The historical approach, with its aim of detecting how things begin and of arriving from these origins at a knowledge of their nature, is certainly perfectly legitimate.... Often it is only after immense intellectual effort, which may have continued over centuries, that humanity at last succeeds in achieving knowledge of a concept in its pure form. (Frege 1884, p. vii)

He then did not believe that knowledge sprouts in the individual mind like leaves on a tree, or is to be accounted simply and solely by reference to the genius of a single man. On the contrary, he suggested that knowledge is typically the result of a continual process of inquiry. Thus he acknowledged that the symbolism for logical relations originated with Leibniz and was revived in modern times by Boole, Grassmann, Jevons, and others who brought out the logical forms to which he then added content (Frege 1882b, p. 88). Hence, whilst admitting that logic is an arena for revolutions and that Frege performed one, it turns out however that such a revolution would not have occurred if the mid-nineteenth century had not prepared the ground for a revolution in logic.

This revolution emanated from mathematics. Frege, whose mathematical specialty like Boole's, had actually been calculus, was seeking to provide mathematics with a solid foundation. It was whilst working in this direction that he stumbled upon the imperfections of natural language, and whilst attempting to overcome these obstacles by his conceptual notation, he switched from mathematics to logic. Boole and Frege's work presupposed on the one hand the generalisation of algebra, and on the other hand the generalisation of the theory of functions. They both relied upon the development of mathematics. As Boolos points out, 'the *Begriffsschrift* contains substantial mathematical results on the theory of relations which must be considered significant contributions to the abstract or generalizing tendency in mathematics that began in the late nineteenth century' (Boolos 1998, pp. 135–136).

It was therefore the development of mathematics which stimulated the regeneration of logic. Boole was particularly distinguished by generalising and systematising syllogistic inferences, and thereby conceived logic as algebra. As was shown earlier, in 1847 he even anticipated the truth-value analysis of 'logical connectives' in the *Mathematical Analysis of Logic* (see subsection **3.5.1**).

Other significant developments were- (i) in papers from the 1880s, Peirce hinted at the possibility of expressing Boole's 'elective function' by only one primitive sign namely 'neither ...nor...', which Sheffer used in his axiomatisation of Boolean algebras. He carried out a refinement of Boole's algebra of logic and, especially, the development of techniques for handling relations within that algebra. He invented quantification theory, and this in complete ignorance of Frege's work (see section **7.1**). According to lecture notes taken by a student of Peirce's, Allan Marquand, Peirce gave a lecture in December of 1879 in which he presented an axiomatisation of arithmetic.[17] The review of this axiomatisation leads Boolos to claim that 'it is certain that the idea of applying the logic of relations to the 'primitive' relation of one number's succeeding another in order to characterize the natural number series was in the air over Baltimore, far from that over Jena, the year the *Begriffsschrift* was published' (Boolos 1998, p. 248).

In 1877 McColl[18] suggested separating the calculus of propositions from the calculus of classes, and then introduced implication as a connective symbol to treat the former. In his paper 'Symbolic Reasoning' he read the expression $a:b$ 'a implies b', or 'if a is true, b must be true'. He called the expression 'implication' or 'conditional statement' which is also 'a very simple and suggestive representation of a great and fundamental laws which runs through all reasoning', namely, *the law of implication*. According to him, 'this law expresses the broad fact that the sole function of the reason is to evolve fresh knowledge from the antecedent knowledge already laid up in the store-house of the memory, and unless we supply it with this material to work upon, it will not work at all' (McColl 1880, p. 52). Thus, two years before the *Begriffsschrift*, he brought forth the first purely symbolical presentation of a variant version of propositional logic which he called 'calculus of equivalent statement'. As Church acknowledges it, 'a true propositional calculus perhaps first appeared...in the work of Hugh McColl...' (Church 1956, p. 156).

In *Was Sind und Was Sollen die Zahlen?* published in 1888, Dedekind showed the validity of the principle of mathematical induction. He provided a definition of the ancestral of a relation. Indeed when Boolos recently read the first draft[19] of *Was Sind und Was Sollen die Zahlen?* he realised that it is quite

---

[17] See the introduction to volume 4 of the *Writings of Charles S. Pierce* p. xliv.

[18] Hugh McColl (1857-1909) wrote several texts in mathematical logic: "The Calculus of Equivalent Statements and Integration Limits," *Proceedings of the London Mathematical Society*, 1877-1878; "Symbolic Reasoning," *Mind*, London, 1880); *Symbolic Logic and its Applications*, London, 1906.

[19] In 1976 the publication of Pierre Dugac's *Richard Dedekind et les Fondements des Mathématiques* contains a large number of previously unpublished texts, including the first draft of *Was Sind und Was Sollen die Zahlen?* mentioned by Dedekind. According to Boolos,

possible that Dedekind formulated the definition for the first time several years before 1878, possibly towards 1872, and quite possibly before Frege arrived at the definition of 'y follows x in the f-sequence' (Boolos 1998, p. 250).

It follows that the line of argument in *Begriffsschrift* was not altogether unique. There was a logical *Zeitgeist* manifest in the work of logicians other than Frege at this time. Frege himself was acquainted with some of these works during the mid-nineteenth century, in particular Boole's logical calculus. Accordingly, in opposition to Dummett, it can be said that Frege was not sailing a sea on which other ships were not in sight.

Sluga too claims that:

> Frege's logic effectively brought to an end the dominance of Aristotelian logic which had been taken for granted in the schools for more than two thousand years. Post-Aristotelian logic begins only with Frege. By terminating the life span of Aristotle's system Frege completed a process that had begun centuries earlier with Galileo's destruction of Aristotelian physics. In the field of logic, it was an epoch-making. (Sluga 1980, p. 65)

Although *Begriffsschrift* propelled a logical renaissance which extended Aristotelian logic, there is again here an exaggeration. Post-Aristotelian logic did not pop out of a *tabula rasa*. It was rather the result of a process of revival of logic, which allowed Boole to introduce mathematics in logic, and Frege to bring logic in mathematics. It is then problematical to fix a definite point for which it may be said (with Sluga): here begins Post-Aristotelian logic. Furthermore, the comparison with Galileo would have led Sluga to mitigate his strong claim. Indeed, in regarding nature as written with mathematical characters, Galileo deciphered this language and undermined Aristotelian physics. But similarly, before Frege, Boole saw that thought can be written in mathematical texture, and started first to point out the fundamental laws of this mathematics of thought. Hence, strictly speaking, it pertains to Boole to be considered as the 'Galileo of thought'.

In addition, in his *Survey of Symbolic Logic* of 1918, C. I. Lewis describes in great details the research programme outlined by Boole, and, at the end of the book, he relegates the logic of *Principia Mathematica* to a brief section of six pages. The first chapter of the book contains the first history of the development of symbolic logic. However, the work of Frege is barely mentioned there. It is only in the last two pages that Lewis refers to Frege, remarking that 'the work of Frege, though intrinsically important, has its historical interest

---

the draft contains most of the ideas and proofs found in the later version (Boolos 1998, pp. 249–250).

largely through its influence upon Mr. Bertrand Russell' (Lewis 1960, p. 114). This shows that logic did not have to wait for Frege to develop, and those who conceive of such a development as entirely dependent upon the work of Frege are historically wrong. Lewis' *Survey of Symbolic Logic* demonstrates indeed what logic looked like to someone who was very familiar with logic.

The exaggerations of the importance of Frege's work have also prompted Putnam to reconsider the account of the history of logic. In an essay, "Peirce the logician," published in his recent collection of articles called *Realism with a Human Face*, he draws attention to the view of Quine according to which 'logic is an old subject and since 1879 it has been a great one'. He takes this claim as a slight to Boole, and stands in opposition to the predominant view that 1879 was an 'epochal' year in the history of logic. He shows that independent of Frege, Peirce had set up a logical language adequate for general logic, which is structurally equivalent to the modern systems. Then, he claims that the possibility of the logical definition of number and the subsequent project of laying down the foundation of arithmetic as an extended logic were independently suggested by Peirce.

Such a tendency to reconsider the place of *Begriffsschrift* in the history of logic is pursued by Boolos who reexamines the view about the history of logic, which he had held for a long time: 'that 1879 was a watershed year for logic'. In '1879?' an essay in his collection of articles called *Logic, Logic and Logic* published in 1998, he begins to sharpen the point by discussing an observation about propositional logic which Boole made towards the end of *The Mathematical Analysis of Logic*. By reference to the section of this booklet called 'Properties of Elective Functions' and to Boole's law of development of functions, he argues that 'Boole clearly had the idea of all possible distinctions of truth-values' (Boolos 1998, p. 245). He then goes on to show how Peirce and Dedekind accomplished remarkable achievements towards the logical definition of number.

The main point of '1879?' is apposite: tremendous achievements in logic had been previously carried out independently of Frege by mathematicians and logicians such as Boole, Peirce, Dedekind. Nonetheless, it should be emphasised that there has never been any question of contesting the importance of the place of *Begriffsschrift* in the history of logic.

Chapter 6

Boole–as–Frege–Discusses–Him

The growth of knowledge depends entirely on the existence of disagreement.

Karl Popper

Introduction

Frege's discussion of Boole's algebra of logic occurs in the following essays: 'Boole's Logical Calculus and the Concept-Script' (1880/81), 'Boole's Logical Formula-Language and my Concept-Script' (1882), and 'On the Aim of the Conceptual Notation' (1882). In the first essay, he spells out the aims and properties of his own symbolic notation, and compares them to Boole's logical calculus, so as to show its 'more far-reaching aim'. The second is a shorter version of the first essay. In the third, Frege discusses Boole's formula-language, analysing it, and pointing out the reason why he does not employ it for his own aim. He brings out two shortcomings of Boole's logical calculus: the first is the fact that the principles of Aristotelian logic have not been altered because of the nonexistence of a theory of quantification; and secondly the fact that Boole uses the same symbols to represent both his 'primary' and 'secondary' propositions', leaving them unrelated to each other.

These three essays are concerned specifically with Boole's logical calculus. They are written between 1879 and 1882 when Frege had to refute the objections to his *Begriffsschrift* raised by Schröder, who claimed that his notation did not differ essentially from Boole's formula-language and could actually be considered as a transcription of the Boole's notation, which would be preferable (Frege 1882c, p. 90).

Given that Frege was performing a strenuous exercise to explain and defend his new logic and the critical tone of these essays, it could be easily thought that he was adjudicating between the two notations, and simply jettisoning Boole's logical calculus. In effect, various Fregean readers have highlighted above all the singularity of Frege's work and have radically insulated it from Boole's. Amongst them Sluga is particularly notable, advancing the standard account of Frege's discussion of Boole, against which I shall take a stand so as to draw a different picture of Boole-as-Frege-discusses-him.

I shall argue that there is a logical lineage between Boole and Frege which would make it untenable to separate completely the Fregean and the Boolean research programmes. Indeed, despite the difference, there is a substantial overlap in content between the two research programmes. Whilst focussing on Frege's essays where he approaches the relation between the two notations, I shall point out that what makes Frege's discussion of Boole important is not only what it tells us about how Frege differentiates himself from Boole, but also what it tells us about how Frege carries further Boole's project of introducing mathematics in logic.

In the essay on "Boole's Logical Calculus and the Concept-Script," Frege writes,

> Despite all differences in our further aims, it is evident from what has been said already that the first problem for Boole and me was the same: the perspicuous representation of logical relations by means of written signs. This implies the possibility of comparing the two. If I now turn to this, it cannot be done in the sense of adjudicating between the two formula-languages, which is to be preferred. To raise such a question would mean referring back to their ultimate aims, which are more ambitious in my case than in Boole's. It would indeed be more than possible that each set of signs was the more appropriate for its own ends. Nevertheless, it seems to me worthwhile to work out the comparison in detail, since in that way many of the peculiarities of my concept-script are thrown into sharper focus. (Frege 1880/81, p. 14)

This passage itself sums up the manner in which Frege conducts his discussion of Boole, and is symptomatic of his legendary intellectual honesty. Frege clearly refuses to adjudicate between the two logical systems, and to decide which is preferable. Instead, he emphasises upon the difference between the two purposes, and acknowledges that it may be plausible that each system would be appropriate for the aim for which it is intended. In 'On the Aim of the Conceptual Notation', he raises the question whether his formal language governs a smaller region than Boole's: 'if the same department of knowledge is symbolized by means of two symbol systems, then it follows necessarily that a translation or transcription from one into the other would be possible.... We can ask whether this translation is feasible throughout, or whether perhaps my formal language governs a smaller region' (Frege 1882c, p. 98).

In order to draw a picture of Boole-as-Frege-discusses-him, I shall regard the above passage as framing such a picture. Frege shares, indeed, the same subject-matter as Boole, though he treats it in a different way. But, this difference of treatment of the same problem is mainly due to the fact that they do not have the same purpose. Boole intends to describe a type of algebraic structure as part of his algebra of logic that allows him to express logical relations as algebraic equations, which are then worked out according to algebraic laws. Whereas what Frege has in mind is the expression of a content that prompts him to set up a notation for logical relations which is suitable for incorporation into the formula-language of mathematics.

I shall consider as a nodal point of Frege's discussion of Boole the thesis that 'the logically primitive activity' is judgement rather than concepts. It is indeed what allows Frege to achieve more far-reaching goals than Boole in yielding the functional analysis of judgements, from which stems the representation of generality that extends beyond Boole's logical calculus.

6.1 The Standard Account of The Discussion

According to Sluga, Heijenoort's insight in his paper, 'Logic as Calculus and Logic as Language', was to argue that, in order to understand the conditions under which metamathematics emerged, the Fregean (or logicist) and the Boolean (or algebraist) traditions must be sharply distinguished in the early history of symbolic logic (Sluga 1987, p. 81). Thus Sluga looks at Frege's own assessment of the relations between his logic and Boolean algebra, in order to extend and adjust Heijenoort's insight.

I shall discuss the account of Frege's discussion of Boole, which Sluga gives in his paper, "Frege Against the Booleans," whose title itself reveals his antagonistic account of the relations between the Boolean and the Fregean research programmes.

Sluga builds his account upon the distinction between *calculus ratiocinator* and *lingua characterica*. He scrutinises the relation between Boole and Frege through what he regards, in a subjective way, as a personal rivalry between

Schröder and Frege.[1] In his review of *Begriffsschrift*, Schröder claimed that 'Frege's title does not correspond at all to the content. Instead of leaning toward a universal characteristic, the present work definitely leans towards Leibniz's *calculus ratiocinator*' (Sluga 1987, p. 83). But, for Sluga 'in contrast, Frege argued that the situation was just the other way around. Boolean algebra was a mere abstract logic, a mere calculus, whereas his own was capable of expressing an actual content and could be considered a partial realization of a characteristic language' (Sluga 1987, p. 83).

Sluga distinguishes sharply between calculus and characteristic language, and expresses the disagreement between the Booleans and Frege in Leibnizian terms, namely, those of *calculus ratiocinator* and of *lingua characterica*. According to him, on the one hand, the Booleans took up one aspect of Leibniz's conception of the logical symbolism, that is, a mechanical calculating procedure. On the other hand, Frege set up a characteristic language whose construction requires a conceptual and philosophical analysis. Then, Sluga draws his picture of Frege's assessment of Boolean algebra as follows:

> If Frege did not consider Boolean logic a *lingua characterica*, i.e. an appropriate notation, that was, in the end, due to the fact he did not believe that the Boolean system of notation gives a proper intuitive representation of the forms of thought and of the structure of human knowledge. A proper logic would, in Frege's eyes, have to be built on the priority principle, reflect the primacy of propositional logic over class logic, show that logic is the foundation of arithmetic, and facilitate the integration of various kinds of knowledge into one symbolism with a single interpretation. (Sluga 1987, p. 92)

There is no doubt that Sluga's 'Frege Against the Booleans' encompasses valuable insights. The recognition of Frege's principle of priority as constituting 'the true center of his critique of Boolean logic' (Sluga 1987, p. 86) is certainly a clear perception of the problem, which I shall accept. He correctly argues that this principle, which gives judgements logical priority over concepts, guided Frege at two points in the construction of his logic which differs from Boole's. One was in the discovery of the function-argument analysis of judgements and the other was in his analysis of general propositions (Sluga 1987, p. 88). However, in attempting to dichotomize sharply the two research programmes through the distinction between a calculus and a characteristic language, Sluga appears to have overdrawn his picture.

---

[1] In fact, Frege considers Schröder's review of his monograph as friendly and attempts to supplement and correct the comparison he makes between Boolean logic and his *Begriffsschrift* (Frege 1880/81, p. 11).

It should be noted that although Sluga clearly heralds that he wants to look at Frege's own assessment of the relation between his logic and Boolean algebra, he does not refer directly to what Frege actually thinks about this relation. Instead, it is through Schröder's review of *Begriffsschrift* that he mostly approaches the relation between Boole and Frege. His use of the term 'assessment' is also inappropriate, in so far as Frege makes clear in his discussion of Boole that he does not attempt to adjudicate between the two logical systems which is to be preferred.

In Frege's conception an adequate notation requires one to supplement the signs of mathematics with a formal element and Boole's logical calculus was unsuited to this task. But this cannot be an objection to Boole since, as Frege recognises, such a supplementation does not enter into his intentions. Hence, in order to prevent any misjudgement between their two formula-languages, Frege warns that whilst comparing them it is always necessary to bear in mind the difference between the purpose that governed Boole in his symbolic logic and the one that governed him in his *Begriffsschrift*.

It is worth returning to the distinction between *calculus ratiocinator* and *lingua characterica*, which Sluga points out as playing a cental role in the 'quarrel' between Frege and the Booleans. A *lingua characterica* is an ideography language in which the logical structure of the expressions follows the structure of the things represented. Such a language is characterised by the use of predicate letters, variables, and quantifiers, and thus is articulated and can express a meaning. It provides a method for setting down the logical relations involved in all scientific knowledge in a written form which is as perspicuous as mathematical notation. A *calculus ratiocinator* by contrast denotes a formal system for reasoning about the logical relations. It is a form of a computation for determining the truth of our propositions, a production of relations through the transformation of formulas according to determinate rules.

On this issue Sluga lines up with Heijenoort, who characterised the difference between the 'Boolean and the Fregean traditions' by drawing attention to the fact that Frege called Boolean algebra a *calculus ratiocinator* whilst describing his own as a *lingua characterica*. In the book, *Frege and Gödel, Two Fundamental Texts in Mathematical Logic*, Heijenoort reiterated the same claim in holding that 'in distinguishing his work from that of his predecessors and contemporaries, Frege repeatedly opposes a *lingua characterica* to a *calculus ratiocinator*' (Heijenoort 1970, p. 3). So Heijenoort and Sluga both want to confine logic to only a *lingua characterica* but misunderstand this as the scope of Fregean logic.

However, Frege was very clear about the relation between *calculus ratiocinator* and *lingua characterica*; he writes,

> I wished to produce, not a *mere calculus ratiocinator*, but a *lingua characterica* in the Leibnizian sense. In doing so, however, *I recognize that deductive calculus is a necessary part of a conceptual notation*.[2] (Frege 1882c, p. 91)

In recognising that 'deductive calculus is a necessary part of a conceptual notation', Frege shows the genuine lineage between his logical system and Boole's. He reiterates his integrative view of the relation between *calculus ratiocinator* and *lingua characterica* in 'On Mr. Peano's Conceptual Notation and My Own,' an essay written in 1897. In it, Frege portrays his achievement as follows,

> In Leibnizian terminology we can say: Boole's logic is a *calculus ratiocinator* but not a *lingua characterica*; Peano's mathematical logic is in the main a *lingua characterica* and at the same time also a *calculus ratiocinator*, whereas my conceptual notation is both, with equal emphasis. (Frege 1897b, p. 242)

Indeed, in Frege's conceptual notation inference is conducted like a calculation in the sense that there is an algorithm there, that is 'a totality of rules which govern the transition from one sentence or from two sentences to a new one in such a way that nothing happens except in conformity with these rules' (Frege 1897b, p. 237). Thus, the logical project of Leibniz, which was seen as mathematics under the two aspects of a *calculus ratiocinator* and of a *lingua characterica*, forms the whole of Fregean logic. And since Boole attempted the first aspect, that is, a *calculus ratiocinator*, it follows that his attempt leads up to one part of Fregean logic.

Accordingly, the lineage between Boole and Frege lies in the fact that they are both concerning with a formal method. Frege's formula language appears to be formal in that the choice of its symbols and the inference derived from these symbols are governed by few explicit rules, in such a way that the conclusion of reasoning is only drawn according to the symbols. This resembles Boole's 'process of analysis', a process by which combinations of interpretable symbols are carried out according to well-determined rules of combination. As Frege acknowledges it, they have at the outset the same problem: 'the perspicuous representation of logical relation by means of written signs' (Frege 1880/81, p. 14).

---

[2] Author's emphasis.

Sluga seems to have perceived the relation between the two systems when referring to Boole's mechanical procedure performed on algebraic symbols. He holds that 'Frege did not want to dispute the significance of such technical improvements in logic. On the contrary, he conceived of his own system as also providing such a technique for problem-solving, such a calculus' (Sluga 1987, p. 84). Indeed, it is actually a calculus of this kind that shows the close relation between Boole and Frege, regardless of the fact that it was carried out in two different ways. For Frege grants that, in Boole's logical calculus, conclusions may be drawn from premises by means of mechanical calculating procedure, even though he is convinced that a complete logic requires more.

Consequently, a fair account of Frege's discussion of Boole should not lead one to oppose radically their two formula languages. Frege is not against Boole. Rather he takes further what Boole started. Their logical systems constitute two overlapping research programmes.

But Boole wants to treat logic as algebra whereas Frege regards logic as fundamental.

6.2 Boole and Frege: The Same Subject–Matter

In the essay 'Boole's Logical Calculus and the Concept–Script', Frege introduces the discussion of Boole by giving credit to Leibniz, who sowed such a profusion of seeds of ideas in the ground, which then were further developed and brought to fruition. Frege counts amidst these seeds the idea of a *lingua characterica* having a closest possible links with that of a *calculus ratiocinator*. Then, he recalls,

> In a short monograph, I have now attempted a fresh approach to the Leibnizian idea of a *lingua characterica*. In so doing, I had to treat in part the same subject-matter as Boole, even if in a different way. (Frege 1880/81, p. 10)

According to Frege, this treatment of the same subject-matter as Boole led Schröder to draw a comparison between Boole's logical calculus and his *Begriffsschrift*, which he attempts now to supplement and correct in the paper.

I shall be stressing the domain common to Boole and Frege's formal languages, that is,'the perspicuous representation of logical relations' and the deductive calculus of reasoning.

6.2.1 The Perspicuous Representation of Logical Relations

Boole and Frege share the same subject-matter: they want to provide a basis for a formal theory of inference. In doing so, they have to construct a symbolical language suitable for replacing ordinary language, which is ambiguous, because of the imperfect correspondence between the disposition of words and the structure of the concepts. They have a common problem, that is, the attempt to dissipate such an ambiguity within the framework of ordinary language. They bring forth an important change, that is, the move from the oral utterance, which indicates imperfectly what a symbolical notation should express clearly, to a perspicuous formulation of logical relations. Thus, in the discussion of Boole's logical calculus, Frege acknowledges that

> It is evident that the first problem of Boole and me was the same: The perspicuous representation of logical relations by means of written signs. (Frege 1880/81, p. 14)

The first step of this move had been made by Aristotle, whose syllogisms already employed schematic letters in place of words. This reliance upon letters instead of words in the process of reasoning began formal logic. There is a great advantage of symbolic notation over speech, which lies in the fact that it clarifies the structure of the relevant arguments. As Leibniz puts it, it 'would literally speak to eyes' (Leibniz 1765, p. 399) rather than to ears, and thus would set up the requirements of a scientific calculation, thereby allowing formal logic, after the discovery of variables, to make a second step by turning reasoning into calculus. If logic is about valid reasoning and the formal rules which govern it, then what is logical must be abstracted from the ordinary concatenation of words, which tend to conceal the actual structure of a proposition. Therefore, any attachment to traditional grammar, which fails to capture the logical structure of a proposition, must be abandoned.

This is what Boole does through the construction of a logical language, in order to express logical relations perspicuously. Hence, Frege, knowing the importance of the necessary departure of logic from ordinary language, acknowledges that it is useful to be acquainted with a means of expression of a quite different kind, such as the formula-language of algebra (Frege 1879/1891, p. 6). And if a *lingua characterica* seeks to depict the structure of a proposition, it is clear then that 'the formula- language of mathematics comes much closer to this goal' (Frege 1880/81, p. 13). Thus, Frege is convinced that the use of arithmetical signs for logical purposes spares us the necessity of learning a completely new algorithm (Frege 1880/81, p. 12). He stresses this conviction again in a letter addressed to Hilbert in which he says that 'the advantages of perspicuity and precision are so great that many investigations could not even have been made without a mathematical sign language' (Frege 1895c, p. 33).

However, it should be acknowledged that Boole's formula-language does not achieve completely the perspicuous representation of logical relations. For, as Frege sees it,

> Anyone demanding the closest possible agreement between the relations of the signs and the relations of the things themselves will always feel it to be back to front when logic, whose concern is correct thinking and which is also the foundation of arithmetic, borrows its signs from arithmetic. To such person it will seem more appropriate to develop for logic its own signs, derived from the nature of logic itself; we can then go on to use them throughout the other sciences wherever it is question of preserving the formal validity of a chain of inference. (Frege 1880/81, p. 12)

Boole's logical language, which borrows its signs from arithmetic, cannot fulfil these high demands, in so far as the language of mathematics, upon which he relies, was still using verbal language in the proof itself, and thus undermining the formal rigour of the inference. Hence, it remains to supplement the signs of mathematics with a formal element. But, according to Frege, Boole's logical calculus, which solves only one part of the problem of the perspicuous representation of logical relations, cannot do that. For it uses the signs $+$, 0, and 1 in a sense which is different from their arithmetical ones. However, 'it would lead to great inconvenience if the same signs were to occur in one formula with different meanings' (Frege 1880/81, p. 13). What then should be done is to develop distinctive signs for logical relations, which can be incorporated into the formula-language of mathematics, and thus to form a complete logical language capable of expressing inferences and definitions involved in mathematics in particular, and science in general. It follows that the problem of the perspicuous representation of logical relations is only completely solved with Frege's concept-script, in which the written formula represents the structure of thought as an objective entity, expressing perspicuously its logical content.

But Frege does not reproach Boole's formula-language its inadequacy of preserving the formal validity of a chain of inference, because, as he acknowledges, it is not Boole's intention to set up his notation for such an application. Then, in referring to logical laws that assume the form of a computation, he admits that 'these means fulfill their purpose, at least as far as the range of problems that Boole has in mind are concerned' (Frege 1880/81, p. 12).

6.2.2 The Deductive Calculus of Reasoning

In 'On The Aim of Conceptual Notation', Frege points out something he has in common with Boole's logical calculus: 'the subordination of concepts'. Indeed, the subordination of one concept to another can be captured in Boole's notation as follows:

$$x = xy$$

For example, if x means the extension of the concept 'men' and y means the extension of the concept 'mortals', then the equation says: the extensions of the concepts 'men' and 'mortals' are the same; that is, 'All men are mortals'.

In opposition to Schröder, who holds that his concept-script has almost nothing in common with Boole's calculus of concepts, Frege shows the contrary in demonstrating how his concept-script can also represent the subordination of concepts. Thus, regarding the judgement

'If $x^2 = 9$, then $x^4 = 81$'

he says: 'now we can call a number whose square is *9* 'a square root of *9*', and one whose fourth power is *81* 'a fourth root of *81*', and translate: all square roots of *9* are fourth roots of *81*' (Frege 1882c, p. 99). Then, as can be readily seen, the concept 'square root of *9*' is subordinated to the concept 'fourth root of *81*'. Unlike Boole's use of 'x', for Frege 'x' expresses the generality of the judgement, which means that whatever we may replace by x, the content should hold. I shall show that his concept-script has a considerable advantage over Boole's notation in that it relies upon a theory of quantification, which allows one to represent the scope of the generality.

In the last part of the essay 'Boole's Logical Calculus and the Concept-Script' (pp. 35–45), Frege also confines his attention to the domain common to the two formal languages. He disregards Boole's primary propositions, and compares their two propositional calculi. As a result, he shows that the axiomatisation of propositional calculus which he sets up in the *Begriffsschrift* needs fewer primitive signs for logical relations, and therefore fewer primitive laws than Boole's logical calculus.

In effect, Frege claims that, in his case, contents of possible judgements expressed by the symbols A and B are connected by the conditional stroke, whereas in Boole's system they are connected by identity, addition and multiplication. Thus, of the four possibilities

A and B
A and not B
not A and B
not A and not B.

Frege observes that Boole's identity sign $A = B$ denies the middle two of the four possibilities. The addition sign $A + B$, which Boole construes exclusively, denies the first and the last; and the multiplication sign AB affirms the first to deny the other three. The inclusive interpretation of the addition sign, which Jevons and Schröder adopt, denies only the last possibility. In order to capture conditionals, Boole employs the notation:

$$A(1 - B) = 0.$$

Thus, there is only two basic signs in Frege to three in Boole. Hence, there exists an inflation of signs in Boole which entails an inflation of primitive rules for computation. This is what strikes Frege, who finds its reason 'in the desire to force on logic signs borrowed from an alien discipline, instead of taking one's departure from logic itself and its own requirements' (Frege 1882d, p. 48).

Since the more primitive signs you introduce, the more axioms you need, by contrast with Boole, Frege follows the principle:

> It is a basic principle of science to reduce the number of axioms to the fewest possible. Indeed, the essence of explanation lies precisely in the fact that a wide, possibly unsurveyable, manifold is governed by one or a few sentences. The value of an explanation can be directly measured by this condensation and simplification: it is zero if the number of assumptions is as great as the number of facts to be explained. (Frege 1880/81, p. 36)

Thus, in order to arrive at the fewest possible primitive signs he chooses those with the simplest possible meanings, seeing that 'the simpler a content is, the less it says'. For instance, he claims, 'my conditional stroke, which only denies the third of the four cases, says less than Boole's identity sign which denies the second as well. The multiplication sign says even more, because it denies the fourth possibility as well, eliminating all choice. Only the addition sign, like my conditional stroke, excludes only one case, if you adopt Stanley Jevons' improvement...' (Frege 1880/81, p. 36). Yet one can express everything one wants with negation and either disjunction or conjunction. But that would not fit with Boole's algebraic approach which demands equations.

Frege attempts to manage everything with the fewest possible primitive laws. By means of these primitive laws, he provides a definition of the ancestral of a relation, and shows the validity of the principle of mathematical induction. In this way, he dovetails with the principles that guide him in setting up the axiomatisation of propositional calculus.

In this axiomatisation, Frege lays down, in the *Begriffsschrift,* nine axioms and five rules; whereas, according to him, Schröder uses fifteen axioms in his *Operationskreise des Logik-kalkuls*. Then, he counts fourteen primitives propositions (nine axioms plus five rules) in his system, and claims to command a somewhat wider domain than does Schröder with fifteen (Frege 1880/81, p. 39). He eventually simplifies the system, so as to obtain eleven basic propositions, and then may claim,

> I see in this the success of my endeavour to have simple primitive constituents and proofs free from gaps. And so I replace the logical forms which in prose proliferate indefinitely by a few. This seems to me essential if our trains of thought are to be relied on; for only what is finite and determinate can be taken in at once, and the fewer the number of primitive sentences, the more perfect a mastery can we have of them. (Frege 1880/81, p. 39)

However, it should be pointed out that in the later development of Boole's logical calculus, Sheffer and Huntington offered a set of five independent postulates, which assumes only one undefined K-rule, and thus dovetails with Frege's basic principle of science, that is, to reduce the number of axioms to the fewest possible (see section **4.3**). Their set of postulates for Boolean algebras are even more condensed and simpler than the system that Frege presents in the *Begriffsschrift*.

In the last six pages of his essay (pp. 39–45), Frege finishes with his comparison in the domain common to the two formal languages by working out Boole's example 5 (Boole 1854, p. 146) in order to illustrate how his concept-script can be used to solve the logical problems Boole tackles, and even do so with fewer preliminary rules for computation (Frege 1880/81, p. 46). I shall not venture into the long-winded calculations of Frege's technique for problem-solving. It suffices to give some indication of how the calculations go. Frege first translates the individual data into his notation. He then picks out the judgements that must be eliminated, and thus obtains the solution among the remained judgements. As he describes it, 'whereas the dominant procedure in Boole is the unification of different judgements into a single expression, I analyse the data into simple judgements, which are then in part already answers to the questions. I then select from the simple judgements those lending themselves to the eliminations needed, and so arrive at the rest of the answers. These will then contain what we wanted to find out' (Frege 1880/81, p. 45).

But, as the editors of the *Posthumous Writings* point out, Frege's procedure of computation contains 'mistakes', and a 'morass of confusion', which may put the correctness of his solution in jeopardy (Footnote 1, p. 41). Indeed, the way in which Boole works out the problem seems to be more appropriate. It takes him only two pages to bring out the solution of the problem, whereas

Frege devotes five pages to it. Hence, in the case of problem-solving, the one-dimensional notation of Boole's logical calculus has an advantage over the Frege's two-dimensional notation of his concept-script, at least in terms of space and time management.

As Frege himself would expect, Boole's logical calculus is better suited than his concept-script to solve the sort of problems for which it is specifically set up. After all, does not Frege say that 'it would indeed be more than possible that each set of signs was the more appropriate for its own ends'? (Frege 1880/81, p. 14)

It is now in this difference of ends that lies the peculiarity of each way of treating the same subject-matter. This is what should be always borne in mind when comparing Boole's system of notation to Frege's if we are not to go astray.

6.3 The Relation of Logic to Mathematics

The two different ends of Boole and Frege, which justify the different way of treating the same subject-matter, bring forth as a corollary two different conceptions of the relation of logic to mathematics. I shall recall their different purposes, and stress that although they do not have the same conception of the relation of logic to mathematics, the works of Boole and Frege show the close relationship between logic and mathematics, in such a way that the separation of the two ceases to be a sharp one.

6.3.1 Boole: Logic as an Auxiliary Part of Mathematics

The purpose of *The Laws of Thought* is 'the construction of a system or method of Logic upon the basis of an exact summary of the fundamental laws of thought' (Boole 1854, p. 66). Boole applies a strong algebraic apparatus to such informal objects as the 'laws of thought'. He points out the relation between the laws of thought and operations like addition and multiplication. Then, he chooses negation, conjunction, and disjunction as the primitive logical operations, and sets up calculi common to thinking in different areas such as, classes, propositions, probabilities, etc,. By relying upon symbolical algebra, he begins with one or more systems of related operations and articulates a common abstract structure. This leads him to posit a set of axioms which is satisfied by each of the systems. It follows that Boole regards logic not as an abstraction from the actual processes of thought, but as a formal construction, for which an interpretation is then sought.

The purpose of Boole then yields a conception of the relation of logic to mathematics. In the *Mathematical Analysis of Logic*, he believes that we ought to associate mathematics with logic and espouses a general view of their relation. Since mathematics is no longer restricted to magnitude, Boole is convinced that there is a common ground upon which logic and mathematics may meet, and exchange with each other. Hence he holds that mathematics and logic are inseparable.

There is not only a formal correspondence between logic and mathematics, but also a partial identity. On the one hand, symbolical algebra, upon which Boole models his logical calculus, is a general science of relations, in which each relation yields a special formal theory with its own axioms and theorems, and a calculus of which these axioms constitute the rules of operation. On the other hand, the logical form of reasoning is a rigorous and explicit deduction in which conclusions are drawn from premises according to general and formal rules set up a priori, and apart from the content of the relations to be considered. It is clear from this that mathematics comes close to logic. Thus, Boole regards both of them as depending upon 'general principles founded in the very nature of language' and upon a resulting 'agreement in process' (Boole 1854, p. 6). It is in the symbolical notation, which achieves the ideal of formal logic that logic and mathematics may be joined together, may assist each other, and eventually may merge.

But, in Boole's eyes, mathematics even does more, in so far as it provides logic with the rigorous and precise forms which make it perspicuous. In *The Laws of Thought*, he claims that 'the ultimate laws of logic are mathematical in their forms' (Boole 1854, p. 11). Thus, if logic is presented in such a way that it consists of symbols and precise rules of operation upon these symbols, then it ensues that it is an auxiliary part of mathematics. In the end, this seems to be Boole's conception of the relation of logic to mathematics.

Such a conception is somewhat in accordance with that of Hilbert, who 'remarked that symbolic logic could be treated as though it were a branch of elementary number theory' (Kneale 1962, p. 714).

6.3.2 Frege: Mathematics as a Branch of Logic

Unlike Boole's logical calculus, which is restricted to a 'pure logic' that does not express a 'content', Frege's *Begriffsschrift* strives 'to make it possible to present a content when combined with arithmetical and geometrical signs'. In 'Boole's logical Formula-language and my Concept-script', Frege tells us his purpose:

> I wanted to supplement the formula-language of mathematics with signs for logical relations so as to create a concept-script which would make it possible to dispense with words in the course of the proofs, and thus ensure the highest degree of rigour whilst at the same time making the proofs as brief as possible. (Frege 1882d, p. 47)

What Frege intends to do is to introduce new signs for logical relations in order to obtain a genuine formula-language, which then can be used to render mathematical inference more rigorous, and thus to lay down the foundations of arithmetic as an extended logic. His explicit purpose is to complete the formula language of arithmetic in such a way that a mathematical proof no longer requires the use of text in ordinary language.

It follows that Frege conceives his logical system neither as an auxiliary part of mathematics, nor as the result of abstractions from reasoning in particular domains. Moreover, the purpose of Frege contains a philosophical insight, that is, the epistemological question of the status of mathematical truths that is the point of departure of his logical investigations. Are they analytic or synthetic? Such a philosophical question cannot be found in the work of Boole. Indeed, the fundamental idea of Frege is that mathematics is a branch of logic. He then aims to reconstruct the whole of that science by means of logical symbolism.

This reductionist programme is sustained by a conception of the relation of logic to mathematics. As Frege sees it, 'mathematics has a closer ties with logic than does any other disciplines; for almost the entire activity of the mathematician consists in drawing inferences.' Thus, since inferring and defining are subject to logical laws, Frege draws the conclusion that 'logic is of greater importance to mathematics than to any other science' (Frege 1914, p. 203). In 'On Formal Theories of Arithmetic,' he also says that 'no sharp boundary can be drawn between logic and arithmetic. Considered from a scientific point of view, both together constitute a unified science' (Frege 1885, p. 112). Although the dichotomy between the two may be understood for practical considerations, Frege believes that this must not become a breach to the detriment of logic and mathematics. He even reminds logicians that 'they cannot come to know their own discipline thoroughly unless they concern themselves more with mathematics' (Frege 1885, p. 113).

However, under the name of 'formal theory', Frege holds the conception that 'all arithmetical propositions can be derived from definitions alone using purely logical means...' (Frege 1885, p. 112) From this he draws the conclusion that 'there is no such thing as a peculiarly arithmetical mode of inference that cannot be reduced to the general inference-modes of logic' (Frege 1885, p. 113); and the requirement that 'everything arithmetical be reducible to logic by means of definitions' (Frege 1885, p. 114). Consequently, Frege uses logic to provide the foundations of mathematics by deriving all mathematical notions from logical ones.

But is it plausible to give a logical definition to the notions of mathematics, or does mathematics possess a specificity so that its logical reconstruction would be inadequate? Unfortunately, some serious fissures, such as the paradoxes, developed out of Frege's reductionist programme, and endangered the solidity of the enterprise to provide a logical basis for the whole mathematical edifice. Frege holds indeed that propositions about natural numbers are propositions about the extensions of certain concepts. The number seven, for example, is the extension of the concept that applies to and only to the concepts that apply to exactly seven objects. But this theory of extensions is shown to be inconsistent due to Russell's paradox.

It is not necessary or appropriate here to discuss the detailed reactions to Russell's paradox (including Frege's own). Nonetheless it is fair to conclude that they all suggest that there is a deep division between logic and set theory (the theory of extensions of concepts); and thus that even if one can model arithmetic on set theory, one does not thereby reduce arithmetic to logic. As Putnam puts it, the central charge laid against Frege's work is that what they (Frege, Russell and Whitehead) called logic is not logic but 'set theory' (Putnam 1990, p. 259).

As a result, Frege's aim to make plausible the derivation of mathematics from logic ceases to have its first sense. Rather the question is now to find a very general system of symbols upon which the whole edifice of mathematics can be built, and this entails that the problem is no longer of a purely logical nature. In this direction, in "A new Attempt at a Foundation for Arithmetic" (1924/25), Frege himself admits that he has 'to abandon the view that arithmetic does not need to appeal to intuition either in its proofs, understanding by intuition the geometrical source of knowledge, that is, the source from which flow the axioms of geometry' (Frege 1924/25a, p. 278). Subsequently, he claims that the whole of mathematics is ultimately the result of 'geometrical source of knowledge', with, of course, the logical one always required where inferences are drawn.

Frege's failure to derive mathematics from logic should however be qualified. For, in his book, *Frege's Conception of Numbers as Objects* published in 1983, Crispin Wright gives a new interpretation of Frege studies which may save Frege from contradiction. Wright endeavours to show that the method followed by Frege in *The Foundations of Arithmetic* is consistent, and can be used for establishing the Peano Axioms without relying upon classes. Thus, in section xix of the book, he provides proofs of the Peano Axioms (see pp. 154–169). He points out that within pure second-order predicate calculus, the Peano Axioms can be derived from what is called 'Hume's principle', that is, ' the number of Fs is identical with the number of Gs if and only if there is a one-one correspondence between the Fs and the Gs'. For Wright, it is possible 'to define arithmetical concepts in terms of logical ones so that, for any particular (decidable) axiomatisation of number theory, a base class of its theorems have purely logical transcriptions which can be demonstrated to be theorems of logic; and from this base class all theorems of the axiomatisation in question follow by logical means' (Wright 1983, pp. 137-138).

Following the way paved by Wright, Boolos argues that 'the number principle', which is related to Frege's principle about sets, turns out to be consistent. As he claims, 'if one reads *The Foundations of Arithmetic* carefully, one sees that Frege uses the set principle *only* to derive the number principle, 'the number of Fs = the number of Gs if and only if the Fs and the Gs are in one-one correspondence.' After deriving the good number principle from the bad set principle, Frege has nothing more to do with sets. Once he has obtained the number principle, he proceeds to show how to derive arithmetic from it with the aid of nothing other than the system of logic he had set out in the *Begriffsschrift*' (Boolos 1998, p. 151).

Although these successful interpretations of Frege's system make it provable that arithmetic is simply a development of logic, it remains a limited success. Possibly, some parts of mathematics can be reduced to logic but no one has yet shown that all mathematics can be so reduced. Moreover, in the paper, "On Formally Undecidable Propositions of Principia Mathematica and Related Systems I" written in 1931, Gödel's famous incompleteness theorem shows that, in the most comprehensive formal system such as, the system of *Principia Mathematica* and the Zermelo-Fraenkel axiom system of set theory which contain the laws of simple arithmetic, there would always be true propositions of the system that could never be proved or disproved within the system. They are undecidable. Thus, by proving that second-order logic and set theory[3] are destined to remain forever incomplete Gödel shows in effect that for any theory which aims to include arithmetic there are propositions which are not provable within the theory but which can be seen to be truths of arith-

---

[3] It might be objected that to describe set theory as 'incomplete' is not quite right, for it is our attempted formalisations that is incomplete, and an extension of our present axiom system may, eventually, generate a complete set theory. However, according to Gödel, this objection is untenable. For in a lecture, 'Some Basic Theorems on The Foundations of Mathematics and Their Implication', delivered in 1951, he shows mathematics to be 'incompletable' or 'inexhaustible' and absolutely undecidable, 'not just within some particular axiomatic system, but by *any* mathematical proof the human mind can conceive' (Gödel 1951, p. 310).

metic. Hence second-order logic and set theory do not seem to belong to the domain of logic, which must have a complete and effective set of axioms like quantification theory or first-order predicate calculus.

Gödel's theorem has an important bearing on the claims of the Fregean research programme. Indeed, although logic is a fundamental theory since it is presupposed by all the deductive theories, such as mathematics, it does not seem to be a good idea to assert that all mathematics can be reduced to logic, if what Frege, Whitehead and Russell call logic is set theory[4], which belongs to the domain of mathematics, and is, by definition, incomplete at any given time. In truth, mathematics and logic are both fundamental and neither reducible one to another.

Furthermore, as well as the contrast between Russell's logicism and Hilbert's formalism which focused on the question whether logic can be construed as a formal system, or whether its symbols demand to be interpreted, there is a general thesis common to both schools of thought: a mathematical system must be constructed in such a way as to allow the inferences and definitions to be carried out without any reference to the meaning of the axioms, although the expressions and rules of operation have been set up on the basis of the meaning of these axioms. This point of likeness of Russell and Hilbert is what has been already construed as the lineage between Boole and Frege who both aim at the outset to construct a formal process as a support of logical deductions. It shows, in a way, the close relationship between logic and mathematics. Which is more fundamental, logic or mathematics, ceases now in a way to be a good question.

Nevertheless, the failure of Frege's reductionist programme does not devalue the logical system which was designed to carry out such a programme. On the contrary, it still remains all-important in that it brings forth a new logic.

6.4 The Nodal Point of Frege's Discussion of Boole

I shall consider, as the nodal point of Frege's discussion of Boole, the thesis that 'the logically primitive activity' is judgement rather than the formation of concepts. In Frege's eyes, Boole's assumption of logically perfect concepts as ready to hand, and judgments to be drawn from by comparing them via their extensions brings him close to Aristotelian logic.

For in Aristotle, as in Boole, the logically primitive activity is the formation of concepts by abstraction, and judgment and inference enter in through

---

[4]I follow here Quine, for whom set theory does not belong to logic, although pioneers in modern logic like Frege, Peano, and various of their followers, notably Whitehead and Russell, viewed it as logic (Quine 1970, pp. 64–66).

an immediate or indirect comparison of concepts via their extensions (Frege 1880/81, p. 15).

As opposed to this, Frege begins with judgments and allows the formation of concepts to proceed from them. His construction of judgements as prior to concept formation, and the representation of generality in mathematics, which flows from his functional analysis of mathematical judgments are presented in the essay "Boole's logical Calculus and the Concept-script" (pp. 14–20). In these pages, Frege shows how his concept-script commands a wider domain than Boole's formula-language.

6.4.1   Judgements as Prior to Concepts

At first sight, it might appear from the title of his booklet, *Begriffsschrift* or "Conceptual Notation," that Frege regards concepts as the logically prior notion from which judgment may be formed. But, this would be a misunderstanding of the leading principle of Frege's work in logic. In the *Begriffsschrift*, 'judgement' is discussed right at the start and this priority of judgement over concepts is affirmed in his essay, 'Boole's Logical Calculus and the Concept-script':

> I start out from judgements and their contents, and not from concepts.... I only allow the formation of concepts to proceed from judgements. (Frege 1880/81, p. 16)

Towards the end of his career Frege reaffirmed the same point in his 'Notes for Ludwig Darmstaedter' (July 1919):

> I do not begin with concepts and put them together to form a thought or judgement; I come by the parts of a thought by analysing the thought. This marks off my concept-script from the similar inventions of Leibniz and his successors, despite what the name suggests; perhaps it was not a very happy choice on my part. (Frege 1919, p. 253)

In Frege's eyes the primacy of judgement over concepts is a distinctive feature of his logic. Although he does not mention it explicitly in the *Begriffsschrift*, this important principle underlies the creation of his formula-language. It encompasses a fundamental innovation in logic, which no longer starts from concepts so as to build up judgments. Rather, concepts are obtained by splitting up a judgement content into parts.

Consequently, Frege can avoid the division in Boole's logical calculus between primary and secondary propositions that gives the priority to 'primary propositions.' In Boole's logical calculus, 'primary propositions' are treated by the logic of classes, which embraces syllogistic theory, and 'secondary propositions' are the object of propositional calculus. Boole (1854, p. 53) gives as an example of a primary proposition:

(1) Animals are either rational or irrational

and of a secondary proposition:

(2) Either animals are rational or animals are irrational

With x standing for 'rational', y for 'irrational', and z for 'animals,' Boole would express the two propositions respectively as,

$$(1) z = x(1-y) + y(1-x)$$
$$(2) x(1-y) + y(1-x) = 1.$$

In (2), 1 represents the universe of discourse, that is, 'animals' under which everything that is being considered falls. Thus, although he employs the same algebraic notation to represent them, these two propositions have different interpretations; and he fails to identify the logical connection between them. However, as it can be readily seen, the first follows from the second. But, Boole cannot express this in his logical calculus, which is incapable of representing inferences in which both primary and secondary propositions occur. In his paper, 'From Boole to Frege' (1976), Dudman brings out very clearly the failure of Boole's account for the affinity between the propositional calculus and the class calculus. But in what does this affinity consist?[5]

As Dudman says, 'in answering this question Boole aligns himself with his traditional predecessors; for according to him the affinity consists in an underlying *identity* between the two logics: the logic of hypotheticals is really only the logic of categoricals all over again, but under another disguise' (Dudman 1976, p. 119). Indeed, belief in an underlying identity between secondary proposition and primary proposition prevents Boole from representing both at once. Thus he cannot recognise that (2) implies (1) and is unable to exhibit

---

[5] In 1903, Russell formulated the question as follows: 'the symbolic affinity of the propositional calculus and the class calculus is, in fact, something of a snare, and we have to decide which of the two to make fundamental' (Russell 1903, p. 12). The question had been discussed in the late nineteenth century. For instance, according to Russell, in 'The Calculus of Equivalent Statements' McColl contended for the view that implication and proposition are more fundamental (Russell 1903, p. 12). As for Peirce, he writes: 'I was...led to suppose that the whole non-relative logic (the propositional calculus) was derivable from the principles of the ancient syllogistic.... My friend, Professor Schröder, detected the mistake and showed that the distributive formulae...could not be deduced from syllogistic principles. I had myself independently discovered and virtually stated the same thing (1885, p. 173f). Venn also expressed his view on the issue (see Venn 1881, Chapters 8 and 18).

this implication. In Boole's system, as Frege notes it, ' any logical transition from one kind of judgment to the other—which, to be sure, often occurs in actual thinking - is blocked' (Frege 1882c, p. 93).

Yet Boole is correct in noticing that (1) does not imply (2). Thus (1) and (2) are not equivalent. According to him,

> The proposition, 'Animals are either rational or irrational,' is primary. It cannot be resolved into, 'Either animals are rational or animals are irrational,' and it does not therefore express a relation of dependence between the two propositions connected together in the latter disjunctive sentence. (Boole 1854, p. 53)

But, in *The Mathematical Analysis of Logic*, Boole stresses in a footnote that he is not able to agree with some writers who 'regard it as the exclusive office of a conjunction to connect *Propositions*, not *words*' (Boole 1847, p. 59).

However, Frege rejects Boole's claim that it is not the exclusive office of a 'conjunction' to connect propositions. He is able to do so because his logical system is based on singular propositions, such as 'Socrates is mortal' or 'This animal is either rational or irrational'. As Boole himself suggests, when in the predicate in such a singular proposition the connective 'or' occurs as a propositional operator, it is allowable to be rewritten as a 'conjunction' of propositions:

> This animal is *either* rational or irrational, is equivalent to, *Either* this animal is rational, or it is irrational. This peculiarity of *singular* Propositions would almost justify our ranking them, though truly universals, in a separate class. (Boole 1847, p. 59)

Hence there might be a way of rewriting (1) in which it is the office of the connective 'or' to construct propositions out of propositions, as it does in (2) (see Dudman 1976, pp 123–28, for useful elaboration). Frege rejects the traditional and Boolean view of singular as 'truly universals' and considers the singular proposition as the fundamental unit of predication. By doing so, he is able with the help of the device of quantification to bridge the gap between primary and secondary logic which for Boole is unbridgeable.

As a result, Frege provides us with a logic which enables both (1) and (2) to be symbolised in such a way that the same symbols have the same interpretation, and also allows the implication with (2) as antecedent and (1) as consequent to be exhibited in a way that reveals its validity. Indeed, Frege's logical system allows the expression of this implication in modern notation as,

(3) $[(\forall x)(Ax \Rightarrow Rx) \vee (\forall x)(Ax \Rightarrow Ix)] \Rightarrow (\forall x)[Ax \Rightarrow (Rx \vee Ix)]$.

Thus (3) shows how (2) and (1) can be integrated within a single, comprehensive logical structure. The quantifier appears here as the most important discovery of Frege which allows him to show that the difference between primary and secondary logic is merely a difference of scope. Indeed, the quantifier explicitly delimits the distinction between (1), that is the universality of a disjunction, and (2), that is a disjunction of universalities. The symbols '$\Rightarrow$' and '$\vee$' are employed as truth-functional propositional operators. (3) is also a conditional in which (2) is shown to imply (1). But it makes it clear that the converse does not hold. (3) is a theorem of the axiomatic system of the predicate logic which Frege sets up in the *Begriffsschrift*. In effect, by observing that Frege's axiomatic treatment of the logic of truth-functions distributes universal quantification through conditionality, Dudman sets out to demonstrate the theorem with the help of the fundamental axiom of generality and the transformation rules, such as the rule of detachment, the rule of substitution and the rule of confinement (see Dudman 1976, pp. 132–133).

Frege now can answer the question about the affinity between the propositional calculus and the class calculus. Unlike Boole, he succeeds in accounting for that affinity by making secondary logic fundamental. In Frege, as Dudman puts it, 'for the first time, logic is presented as an organic whole, with the predicate calculus *based* on the sentential calculus (Dudman 1976, p. 134). Indeed, Frege's progress over Boole is his construction of the first unified logical system which bases primary logic upon secondary logic. This construction is possible because of his discovery of quantification, 'the deepest single technical advance ever made in logic' (Dummett 1981b, p. xxxiii). Such a discovery is even deeper insofar as it enables Frege to give an account of inferences involving multiple generality which are common place in mathematics.

Moreover, Boole reduces the 'secondary propositions' to the 'primary propositions' in construing the conditional judgment 'if B then A' as a case of subordination of concepts, interpreted as 'the class of time instants at which B is included in the class of time instants at which A'. So he represents the proposition as,

$$A(1 - B) = 0,$$

which is the expression of the general proposition, All As are Bs. Thus, for Boole, 'primary propositions' are prior to 'secondary propositions.' He analyses a general proposition as having a subject and predicate as its necessary structure. For instance, in the proposition, 'Every square root of 4 is a 4th root of 16', 'Every square root of 4' is the subject, and 'a 4th root of 16' is the predicate. His analysis of the proposition would then consist of extracting these two concepts as distinct elements within the proposition, and putting them together so as to form a judgement. In truth, Boole's analysis of propositions in terms of subject and predicate is one reason why his logical calculus is unsuitable to carry out some valid inferences involving generality that occur in mathematics.

On the other hand, Frege claims:[6]

> In contrast with Boole, I now reduce his *primary propositions* to the *secondary* ones. (Frege 1880/81, p. 17)

He considers secondary proposition as prior and the basis of his logical system. Thus, he interprets the proposition, 'Every square root of 4 is a 4th root of 16', or the subordination of the concept 'square root of 4' to the concept '4th root of 16' as a complex one in that it is composed from several 'quasi-sentences' [7]. The proposition means:

> if something is a square root of 4 then it is a 4th root of 16.

Here, there are two 'quasi-sentences,' 'something is a square root of 4' and 'it is a 4th root of 16', which are connected. The connection can be seen in the relationship between 'something' in the first 'quasi-sentence' and 'it' in the second. It is more apparent when the proposition is rewritten as, 'if something is a square root of 4 then that thing is a 4th root of 16.' In order to show readily the connection, let us substitute for 'something' and 'it' the same proper name '2', and obtain a proposition which is a particular case of the above general proposition. Thus, we have, 'if 2 is a square root of 4 then 2 is a 4th root of 16, which is true if the above proposition is true.

If we now recur to Frege's analysis of proposition in terms of function and argument (see subsection **5.3.2**), which is based upon his thesis that judgements are prior to concepts, then we can substitute 'something' and 'it' for a variable 'x'and obtain:

> 'if x is a square root of 4, then x is a 4th root of 16',

in which the functional expressions 'being a square root of 4' and 'being a 4th root of 16' are joined together by the conditional connective. The variable 'x'

---

[6]However, Dudman argues that in Frege's case there is no attempt to *reduce* primary to secondary. He claims that 'on the contrary, when, in BS (*Begriffsschrift*), we move up from the logic of hypotheticals to the logic of categoricals, there swing into action for the first time one new primitive (the quantifier), one new axiom (§22) and one new transformation rule (the Confinement Principle). Quantification theory does not reduce to the logic of truth functions but is superimposed upon it (1976, p. 134).

[7]Frege calls something 'a *quasi-sentence* if it has the grammatical form of a sentence and yet is not an expression of a thought, although it may be part of a sentence that does express a thought, and thus part of a sentence proper' (Frege1906, p. 190).

serves for the expression of the generality of the proposition.[8] Hence Frege reduces Boole's primary proposition, 'All As are Bs' to the complex conditional proposition, 'if A then B'. The reduction proceeds from terms (A/B) to predicates $(A(x), B(x))$ and with the introduction of the universal quantifier All As are Bs is construed as $(\forall x)(A(x) \Rightarrow B(x))$. Thus, the above expression can be written as,

'For all x if x is a square root of 4, then x is a 4th root of 16'.

The introduction of the quantifier 'For all x' enables Frege to express the scope of the generality. As a result, he claims: 'I believe that in this way I have set up a simple and appropriate organic relation between Boole's two parts' (Frege 1880/81, p. 18). Indeed, Boole regards the two parts of his logical calculus as two different interpretations of the same algebraic structure, whereas Frege sets out to 'give a homogenous presentation of the lot' (Frege 1880/81, p. 14).

It follows that the thesis of the priority of judgements over concepts, which lightens his way of analysing propositions, allows Frege to carry out a new logic. It leads him to two points in the setting up of his logical system: the functional analysis of mathematical judgements and the representation of generality in mathematics. I shall consider these two implications of the priority thesis within Frege's discussion of Boole.

6.4.2 Functional Analysis of Mathematical Judgments

In the discussion of Boole, in order to illustrate how the method of extraction of function works, Frege gives as an example this mathematical possible judgement:

$$2^4 = 16.$$

The analysis in terms of subject-predicate would suggest that this possible judgment has a single, unique structure, whereas the method of extraction of function shows that the same content, i.e., $2^4 = 16$ can be analysed in various way so as to obtain the above possible functions:

$$x^4 = 16; \quad 2^x = 16; \quad x^4 = y.$$

For instance, since Frege allows the formation of concepts to proceed from judgments, we may imagine the 2 in the content of the possible judgment

$$2^4 = 16;$$

---

[8]I use here the conventional variable 'x', but Frege himself prefers to use the italic letters, 'a', 'b', for the reasons indicated in section **5.3.1**.

to be replaceable by (-2) or by 3, which may be indicated by putting an x in the place of the 2:
$$x^4 = 16.$$
The content is thus split into a constant and a variable part. 'The former, regarded in its own right but holding a place open for the latter, gives the concept '4th root of 16'.

Thus, Frege expresses
$$2^4 = 16$$
by the sentences '2 is fourth root of 16' or the individual 2 falls under the concept '4th root of 16'. But it is also possible to say that '4 is a logarithm of 16 to the base 2'. The 4 is taken to be replaceable so as to obtain the concept 'logarithm of 16 to the base 2':
$$2^x = 16$$
with the x standing for the sign for the individual falling under the concept. Here the 16 in $x^4 = 16$ may also be regarded as replaceable in its turn by
$$x^4 = y$$
which expresses the relation of a number to its 4th power. Accordingly, Frege claims:

> Instead of putting a judgment together out of an individual as subject and an already previously formed concept as predicate, we do the opposite and arrive at a concept by splitting up the content of possible judgment. (Frege 1880/81, p. 17)

For Frege, if a content of possible judgement is analysable in this way, then it must be already articulated. However, this does not mean that concepts and relations involved are constituted apart from objects. On the contrary, Frege recognises them as given with the first judgement in which they are ascribed to things. Hence, he writes that 'in the concept-script their designations never occur on their own, but always in combinations which express contents of possible judgement.' He then compares this process to the behaviour of the atom: 'we suppose an atom never to be found on its own, but only combined with others moving out of one combination only in order to enter immediately into another' (Frege 1880/81, p. 17). So a concept is never found on its own, but always occurs in the context of a judgement.

Dummett states as a thesis Frege's method of extraction of function:

> Thought is not built up out of its components concepts, rather, the constituents of the thought are arrived at by analysis of it. (Dummett 1981a, p. 261)

This thesis also appears in a letter which Frege addressed to Marty:

> I think of a concept as having arisen by decomposition from a judgeable content. (Frege 1882a, p. 101)

On the other hand, Boole, following the tradition, would take the opposite way: that judgements are drawn from concepts by putting them together in different ways. This is the result of his belief that concepts are given and logically prior over judgements. Hence what has been considered as the nodal point of Boole-as-Frege-discusses-him surfaces here again. Frege's thesis that judgements are prior to concepts leads him to the analysis of generality and constitutes what differentiates him from Boole.

6.4.3 Representation of Generality in Mathematics

In the interpretation of the subordination of the concept 'square root of 4' to the concept '4th root of 16' as meaning: 'if something is a square root of 4 it is a 4th root of 16', Frege claims to distinguish between concept and individual, which Boole does not, for his letters never mean individuals but always extensions of concepts.

In Frege's eyes, there must be a distinction between concept and thing, even in the case in which only one thing falls under a concept. As he puts it, 'the concept "planet whose distance from the sun lies between that of Venus and that of Mars" is still something different from the individual object the Earth, even though it alone falls under the concept' (Frege 1880/81, p. 18). This distinction allows him to form concepts with different contents whose extensions are restricted to this one thing, the Earth. Likewise, he distinguishes the case of one concept being subordinate to another from that of a thing falling under a concept. Thus, in his concept-script the distinction is shown by representing the sentence 'if something is a square root of 4 it is a 4th root of 16' as follows,

$$\vdash \begin{array}{l} x^4 = 16 \\ x^2 = 4 \end{array}$$

and the sentence 'the individual 2 falls under the concept '4th root of 16' as,

$$\vdash\!\!\!\!-\!\!-\!\!-\!\!-\!\!-\!\!-\!\!-\!\!-\!\!-\ 2^4 = 16$$

Frege then represents generality as well as particular and existential propositions, which have only an inadequate expression in Boole's logical calculus. Thus, he intends, on the one hand, to stress that the difference between Boole's logical calculus and his concept-script is closely bound up with their original purpose, and on the other hand that because of his notation for generality, his domain of logic was wider than Boole's.

Indeed, Frege aims to set up his concept-script for mathematicians, so that they can carry out their inferences perspicuously; hence he is concerned with mathematical judgements such as, 'All square roots of 4 are 4th roots of 16'. The generality in this judgement symbolised as,

$$\vdash\!\!\!\!-\!\!-\!\!-\!\!-\!\!-\!\!\top\!\!-\!\!-\ x^4 = 16 \\ \llcorner\!\!-\ x^2 = 4$$

is expressed by the symbol x. The judgment is valid whatever we may replace by x. Frege stipulates that 'the roman letters used in the expression of judgements should always have this sense' (Frege 1880/81, p. 18). The roman or italic letter 'x' is now what is called a free variable. It has as its scope the content of the whole judgement, and this need not be indicated by a quantifier notation (see Frege 1879, p. 131). Then he considers the case where the content of such a general affirmative judgement occurs as an antecedent of a hypothetical judgement; e.g. 'if every square root of 4 is a 4th root of m, then m must be 16'. The content of this judgement may also be rendered as follows: 'if, whatever you understand by x it holds that $x^4 = m$ must be true if $x^2 = 4$, then $m = 16$'. In the expression

$$-\!\!-\!\!-\!\!-\!\!\top\!\!-\!\!-\ m = 16 \\ \llcorner\!\!\top\!\!-\ x^4 = m \\ \llcorner\!\!-\ x^2 = 4$$

which does not represent the judgement, we may see that the generality to be expressed by x must not govern the whole but must be confined to this part:

$$-\!\!-\!\!-\!\!-\!\!\top\!\!-\!\!-\ x^4 = m \\ \llcorner\!\!-\ x^2 = 4$$

Then Frege says, 'I designate this by supplying the content-stroke with a concavity in which I put a gothic letter which also replaces the x':

$$\vdash\!\!\frown\!\!\underset{\mathfrak{a}}{\frown}\!\!\!\!\top\!\!\!\!\!\begin{array}{l} \mathfrak{a}^4 = m \\ \mathfrak{a}^2 = 4 \end{array}$$

Thus, he confines the generality represented by the gothic instead of italic letter to the content in whose content stroke the concavity occurs. The gothic letter stands for what is now called a bound variable. For the occurrence of the gothic letter lies within the scope of the universal quantifier represented by the concavity which occurs only in one part of the judgement. Accordingly, the judgement may now be expressed as follows:

$$\top\!\!\!\!\!\begin{array}{l} m = 16 \\ \underset{\mathfrak{a}}{\frown}\!\!\!\top\!\!\!\!\!\begin{array}{l} \mathfrak{a}^4 = m \\ \mathfrak{a}^2 = 4 \end{array} \end{array}$$

In "On The Aim of Conceptual Notation," Frege then claims:

> I consider this mode of notation one of the most important components of my 'conceptual notation', through which it also has, as a mere presentation of logical forms, a considerable advantage over Boole's mode of notation. In this way, in place of the artificial Boolean elaboration, an organic relation between the primary and the secondary propositions is established. (Frege 1882c, p. 99)

This notation for generality allows also Frege to express particular and existential judgements. Thus, the judgement 'Some 4th roots of 16 are square roots of 4' is represented as

$$\vdash\!\!\frown\!\!\underset{\mathfrak{a}}{\frown}\!\!\!\!\top\!\!\!\!\!\begin{array}{l} \mathfrak{a}^2 = 4 \\ \mathfrak{a}^4 = 16 \end{array}$$

As for the existential judgements such as, 'there is at least one square root of 4', he gives:

$$\vdash\!\!\frown\!\!\underset{\mathfrak{a}}{\frown}\!\!\!\!\top\!\!\!\!\!\begin{array}{l} \mathfrak{a}^2 = 4 \end{array}$$

As a result, Frege says that 'even when we restrict ourselves to pure logic my concept-script commands a somewhat wider domain than Boole's formula-language. This is a result of my having departed further from Aristotelian logic' (Frege 1880/81, p. 15). This departure from Aristotelian logic resides in his treatment of judgement as the logically primitive activity, which has been construed as central in his discussion of Boole.

In Boole's logical calculus and the concept-script (pp. 21–27) since the difference in extent of the domains governed by Boole's logic and his concept-script depends upon their further purposes, Frege shows through examples how 'the construction of his notation enables it, when combined with the signs of arithmetic, to achieve the more far-reaching goals it set itself (Frege 1880/81, p. 21). Thus, he represents complex mathematical propositions which have only been expressed in words. For instance, he gives the definitions of the continuity of a function, of a limit, and that of following in a series, which appears in section 26 of *Begriffsschrift*. Then, Frege proves an arithmetic proposition such as, 'the theorem that the sum of two multiples of number is in its turn a multiple of that number', by giving its formal derivation (pp. 27–32). In all those respects Frege is convinced that Boole's formula-language is unsuited, and therefore his logical system is more ambitious in that it 'sees' further than Boole's.

6.5 A Picture of Boole–as–Frege–Discusses–Him

The point where a picture of Boole–as–Frege–discusses–him may be drawn is now reached. It has been shown that in his comparison between Boole's logical calculus and his concept-script, Frege confines himself to comparing the two formula-languages whilst bearing in mind the difference of their respective purposes. He neither adjudicates between the two formula-languages, so as to decide the more preferable, nor does he simply jettison Boole's logical calculus. Rather, he shows the domain common to the two formal languages, although he stresses his different and more ambitious way of treating it. It follows that what makes Frege's discussion of Boole important is not only what it tells us about how Frege sees his difference from Boole, but also what it tells us about how Frege carries out further Boole's attempt to attach mathematics to logic.

Thus the standard account of Frege's discussion of Boole, which focuses mainly upon what differentiates them, misses thereby the mathematical logic lineage between them, and misapprehends what Frege actually performed: a revolution in logic in which Boole's logical calculus is overcome and subsumed as a calculus of classes. But this needs to be qualified by a proper understanding of the way in which Boole's research programme was developed in modern Boolean algebra of logic. Indeed, an interesting development of Boole's logical calculus can show how closely it is related to Frege's first-order predicate calculus.

In effect, Boole carried out a calculus of classes which was concerned with the relations between classes whose elements, if they are not empty, are all contained in a common domain of individuals called the universe of discourse. In *Basic Laws of Arithmetic*, Frege too developed a calculus of classes which was concerned not only with relations of inclusion, intersection, and disjunction, between classes of the same level, but also with the membership of one class in another. Frege's theory of classes will find its systematic and exhaustive development in Whitehead and Russell's *Principia Mathematica*.

However, it turns out that in his axiomatisation of Boole's logical calculus Huntington showed the connection between Boolean algebra and the *Principia Mathematica* through a comparison between his fourth set and the elementary propositions presented in Section A of the *Principia Mathematica* (see subsection**4.3.1**). This comparison led him to point out the close relation between the two axiomatic systems. In truth, it appears that the modern Boolean algebra of unions, intersections, and complements, which has various interpretations, is equivalent to that part of first-order predicate calculus which employs only one-place predicate letters. For the variables in Boolean algebra are unquantified and can be interpreted as schematic one-place predicate letters. Alternatively, as Kneale sees it, Boole's algebra of classes is a fragment of the restricted calculus of propositional function (Kneale1962, p. 624).

It follows that although historically distinct the Boolean and the Fregean research programmes are not incompatible, and therefore a complete separation of the two is not appropriate. Moreover, after the failure of the Fregean research programme into the foundations of mathematics, systematic work in metamathematics resembles somewhat Boole's method, insofar as it inquires into the validity of well-formed formulas capable of various interpretations. Indeed, Löwenheim's paper of 1915 is a work within both Boole and Frege's research programmes, and somewhat a blending of the two logical calculi. In this connection, it is more than regrettable that the correspondence between Frege and Löwenheim was ultimately lost in the Second World War. For, according to Sluga, in this correspondence, Frege was eventually convinced of the possibility of an investigation of logical and mathematical formalism as an uninterpreted calculus (Sluga 1987, p. 94).

Chapter 7

Metamathematics: A Return to Boole's Research Programme

> *...in playing music, a series of processes which were originally conscious must have become unconscious and mechanical so that the artist, unburdened of these things, can put his heart into the playing.*
>
> G. Frege

Introduction

The concept of 'research programme' has been used as a tool for the analysis of the development of mathematical logic from Boole to Frege. This has led to the study of the two research programmes and the discussions of their relative merits as well as the review of their relationship. As a result, it has been shown that the progress made in mathematical logic stemmed from two continuous and overlapping research programmes: Boole's introduction of mathematics in logic and Frege's introduction of logic in mathematics. It has then been argued that there is a progressive continuity in the development of mathematical logic, and if that science is an arena where revolutions take place, it is not, however, the case that when a previously existing research programme is overthrown, it is irrevocably discarded. Rather, a successor research programme must respect the achievement of the research programme it displaces. Moreover, since it has been claimed that the 'normal' state of logic is not characterised by routines and puzzle-solving, but by the redevelopment and extension of earlier logical research programmes, the concept of 'revolution' has not been used as the core unit for analysing the history of logic.

This line of reasoning dovetails with the historical cartography of mathematical logic drawn by Hilbert and Ackermann. Their mapmaking of the development of that science may be worth reproducing here in full:

> The first clear idea of a mathematical logic was formulated by Leibniz. The first results were obtained by A. Morgan (1806–1876) and G. Boole (1814–1864). The entire later development goes back to Boole. Among his successors, W. S. Jevons (1835–1882) and specially C. S. Peirce (1839-1914) enriched the young science. Ernst Schröder systematically organized and supplemented the various results of his predecessors in his *Vorlesungen über die Algebra der Logik* (1890–1895), which represents a certain completion of the series of developments proceeding from Boole. In part independently of the development of the Boole-Shröder algebra, symbolic logic received a new impetus from the need of mathematics for an exact foundation and strict axiomatic treatment. G. Frege published his *Begriffsschrift* in 1879 and his *Grundgesetze der Arithmetic* in 1893-1903. G. Peano and his co-workers began in 1894 the publication of the *Formulaire des Mathématiques*, in which all the mathematical disciplines were to be presented in terms of the logical calculus. A high point of this development is the appearance of the *Principia of Mathematica* (1910–1913) by A. N. Whitehead and B. Russell. Most recently Hilbert, in a series of papers and university lectures, has used the logical calculus to find a new way of building up mathematics which makes it possible to recognize the consistency of the postulates adopted. The first comprehensive account of these researches has appeared in the *Grundlagen der Mathematik* (1934–1939), by D. Hilbert and P. Bernays. (Hilbert and Ackermann 1928, pp. 1–2)

To this integrative picture of the historicity of mathematical logic, one should add Löwenheim's paper of 1915, Gödel's first incompleteness theorem of 1931, and the crucial work carried out from the 1930s onwards, which is one of the most remarkable events in the development of mathematical logic, bringing together many brilliant minds working on closely related programmes. It is such an integrative picture that underlies this treatment of the development of mathematical logic from Boole to Frege in terms of overlapping research programmes.

In what follows, I shall show the importance of Boole in the history of logic. I stress that Boole's research programme through the work of Peirce developed a propositional calculus and a predicate calculus of functions of one and of several variables with quantification, and Boole himself made a significant contribution to the development of mathematical logic which should not be overlooked. Then I shall inquire about the conditions under which metamathematics emerged. I argue that Boole should earn the right to be considered as the grandfather of metamathematics. The argument is substantiated on the one hand by developing the idea that Boole's formalist treatment of logic leads on to Hilbert's metamathematics, and on the other by regarding Boole's semantics as having suggested the model theoretic approach to logic that is prominent in Löwenheim's paper of 1915, which constitutes a revival of the Boolean research programme. This leads me back to Gillies' 'The Fregean Revolution in Logic'. I suggest that the emergence of the model-theoretic approach to logic should not be regarded as a part of the Fregean revolution, but a distinct research programme whose possibility required the research programme of Boole. Finally, in order to stress the close relationship between Boole and Frege, I draw an overall portrait of them, which was actually portrayed by Frege himself.

7.1 The Importance of Boole in The History of Logic

Boole is one of the greatest logicians since Aristotle. As Peirce holds it, his book, *The Laws of Thought* 'is destined to mark a great epoch in logic; for it contains a conception which in point of fruitfulness will rival that of Aristotle's *Organon*' (Peirce 1865, p. 224). However, the historical and logical significance of his work has been generally overlooked. For instance, in his *Esquisse d'une histoire de la logique*, which gave a decisive impetus to the renewal of historical studies of logic, Scholz reserved only a few lines of allusion to Boole's creation of the calculus of logic (Scholz 1931, pp. 88–89). Furthermore, since the appearance of the theory of quantification, the critics of Boole have considered him from the perspective of a logician of our time. Accordingly, they have tended to assess his logical calculus in accordance with modern standards of rigour. In this direction, Dummett's assessment of Boole has been particularly influential.

Dummett contends:

> Boole cannot correctly be called 'the father of modern logic'. The discoveries which separate modern logic from its precursors are of course the use of quantifiers (or, more generally, of operators which bind variables and can be nested) and the concept of a formal system, both due to Frege and neither present even in embryo in the

> work of Boole. Boole has indeed a great historical importance both
> for abstract algebra and for logic. As had Leibniz two centuries
> earlier, he devised a general theory of classes under Boolean oper-
> ations, a theory which of course contained the traditional theory
> of the syllogism. This move gained its importance for logic rather
> from the novelty of any extension of logical theory than from the
> magnitude of the extension itself; and anyone unacquainted with
> Boole's works will receive an unpleasant surprise when he discovers
> how ill-constructed his theory actually was and how confused his
> explanations of it. (Dummett 1978, p. 67)

That Boole's logical calculus is not rigorously developed compared to modern quantification theory is certainly true. It is certainly true that it cannot symbolise properly statements like 'Some men are vegetarians', or dyadic and higher-degree relations, and that there are no quantifiers in his logical system. But it is altogether not appropriate to assess Boole's work according to modern standards of rigour, overlooking the fact that it has not been carried out in our time.

Dummett gives us the impression that Frege discovered the quantifier *ex nihilo* by himself. But, as we have seen, Frege's work is not the only source for our modern conception of the quantifier. For, within the Boolean research programme, and quite independently of Frege, Peirce, his student O. H. Mitchell and Schröder had developed a logic of relations in which they had presented the quantifier.[1] They introduced a typographical notation that, like the modern one, lends itself to writing formulas on a line in contrast with Frege's notation, which is two-dimensional. It also allows a simple analysis of normal-form formulas into a prefix, which Peirce called the quantifier, and a matrix that he called 'the Boolean part' of the formula.

In the paper, "On the Algebra of Logic: A Contribution to the Philosophy of Notation" published in the *American Journal of Mathematics* in 1885, when Peirce came to the distinction of *some* and *all*, he noted:

> All attempts to introduce this distinction into the Boolean algebra
> were more or less complete failures until Mr. Mitchell showed how
> it was to be effected. His method really consists in making the
> whole expression of the proposition consist of two parts, a pure
> Boolean expression referring to an individual and a Quantifying

---

[1] Peirce credits the idea of quantifier to his student O. H. Mitchell who discovered it four years later than Frege in his publication "On a New Algebra of Logic" in *Studies in Logic* (1883). But the use of an operator variable in connection with the quantifier was a contribution of Peirce as a modification of Mitchell's notation.

Metamathematics: A Return to Boole's Research Programme 225

part saying what individual this is. Thus, if k means 'he is a king,' and h, 'he is happy,' the Boolean

$$(\overline{k} + h)$$

means that the individual spoken of is either not a king or is happy. Now, applying the quantification, we may write

$$\text{Any } (\overline{k} + h)$$

to mean that this is true of any individual in the (limited) universe, or

$$\text{Some } (\overline{k} + h)$$

to mean that an individual exists who is either not a king or is happy. So

$$\text{Some } (kh)$$

means some king is happy, and

$$\text{Any}(kh)$$

means every individual is both a king and happy. (Peirce 1885a, pp. 178–179)

What is clearly noticeable here is the quantification structure. Peirce credited Mitchell for quantifying the notions of 'any' and 'some' in Boolean algebra. By a 'pure Boolean expression' Peirce meant here roughly what we would now call 'a propositional function' and by 'the Quantifying part' what we would now call 'the quantifier'.

But, as Peirce pointed out, 'the algebra of Boole affords a language by which anything may be expressed which can be said without speaking of more than one individual at a time' (Peirce 1885a, p. 177). Thus a Boolean expression can only make an assertion about an individual class. It is not expressive enough to represent relations of considerable complexity as the complicated inferences in mathematics, for polyadic quantifiers would be needed to do so.

In effect, in order to bring out the detailed structure of relational predicates, not only properties of individuals are required, but also binary relations between individuals and some way of handling more than one variable at a time. As will be seen, Peirce focused on a logic of many variables, thereby developing a semantics for first-order predicate logic. He was capable of dealing with polyadic predicate logic, that is, the general case of quantifying an argument of a propositional function of many variables.

In his 1885 paper, Peirce carried out, indeed, a new system of his own which he called 'the first-intentional logic of relatives' (p. 177). The system involves individuals variables and quantifiers and drops relative product and relative sum. Peirce specifically was concerned with the quantifiers as an object of investigation. Thus he introduced his system by explicating the notation he was using for them and what the quantifier operations mean. He wrote:

> Here, in order to render the notation as iconical as possible we may use $\sum$ for *some*, suggesting a sum, and $\prod$ for *all*, suggesting a product. Thus $\sum_i x_i$ means that x is true of some one of the individuals denoted by i or
> $$\sum_i x_i = x_i + x_j + x_k + \ldots$$
> In the same way, $\prod_i x_i$ means that x is true of all these individuals, or
> $$\prod_i x_i = x_i x_i x_k \ldots$$
> If x is a simple relation, $\prod_i \prod_j x_{ij}$ means that every i is in this relation to every j, $\sum_i \prod_j x_{ij}$ that some one i is in this relation to every j, $\prod_j \sum_i x_{ij}$ that to every j some i or other is in this relation, $\sum_i \sum_j x_{ij}$ that some i is in this relation to some j. (Peirce 1885a, p. 180)

Peirce's explication of the meaning of the quantifier operations is clearly an explication not at the level of syntax but at the level of semantics. Moreover, he pointed out the use of individual variable (in the above notation 'x' is a predicate, monadic or dyadic, not an individual variable), which he called 'indices', as an important aspect of the notion of quantifiers and gave their linguistic meaning as follows:

> The index asserts nothing, it only says 'There!' It takes hold of our eyes, as it were, and forcibly directs them to a particular object, and there it stops. Demonstrative and relative pronouns are nearly purely indices, because they denote things without describing them; so are the letters on a geometrical diagram, and the subscript numbers which in algebra distinguish one value from another without saying what those values are. (Peirce 1885a, p. 163)

Metamathematics: A Return to Boole's Research Programme 227

Peirce formed his notation for quantification by means of indices. They serve to 'distinguish one value from another without saying what those values are'. Indices are neither names nor constants; they are more like variables. They carry what is being talked about in a given discourse, i.e., individuals. Thus, although indices are mathematical notation, they are interpreted as pronouns. The sense of the word "denote" is supposed to be given by some interpretation function. Indeed, Peirce understood that he was dealing not merely with an algebra but with language. However, he did not follow the characteristic procedure when setting up a first-order predicate logic. He did not begin, for example, by saying what is a well-formed formula and, to this end, by providing formal definitions of variables, predicate symbols, and quantifiers and specific formation rules. Rather, Peirce engaged in informal semantic exposition in presenting his system. Thus he gave examples borrowed from everyday English language to illustrate his notation and how it is to be interpreted. He said:

> Let l_{ij} mean that i is a lover of j, and b_{ij} that i is a benefactor of j. Then
> $$\prod_i \sum_j l_{ij}\, b_{ij}$$
> means that everything is at once a lover and a benefactor of something; and
> $$\prod_i \sum_j l_{ij}\, b_{ji}$$
> that everything is a lover of a benefactor of itself.

And

> Let g_i mean that i is a griffin, and c_i that i is a chimera, then
> $$\sum_i \prod_j (g_i\, l_{ij} + \bar{c}_j)$$
> means that if there be any chimeras there is some griffin that loves them all. (Peirce 1885a, p. 180)

It can be observed that all quantifiers are clustered together at the left side. Thus Peirce wrote the formulas in prenex form with a matrix within which a disjunction of conjunction of formulas of relational expressions is inserted. In this above notation, Peirce certainly reached something not far from modern quantification theory.

Peirce also considered the procedure in working with his first-order predicate logic (see pp. 182–184). Of the seven methods which he pursued, I shall

here give the two inference rules used for his quantifiers. The first rule states that:
$$\prod_i x_i \cdot \prod_j x_j = \prod_i \prod_j x_i x_j,$$
$$\sum_i x_i \cdot \prod_j x_j = \sum_i \prod_j x_i x_j,$$
$$\sum_i x_i \cdot \sum_j x_j = \sum_i \sum_j x_i x_j.$$

Thus, different expressions with distinct indices can be written together, and all the quantifiers can be brought to the left. According to Peirce, this rule is the most useful on the whole. The second rule states that:
$$\prod_i \prod_j x_{ij} = \prod_j \prod_i x_{ij},$$
$$\sum_i \sum_j x_{ij} = \sum_j \sum_i x_{ij}.$$

Thus the quantifiers of different expressions may be moved relatively to one another without deranging the order of the indices of any one expression.

What has been said so far seems to indicate that Peirce was aware of semantical considerations. The discussion of what he called 'the universe' would show again the Peircian semantical approach. He affirmed:

> I propose to use the term 'universe' to denote that class of individuals *about* which alone the whole discourse is understood to run. The universe, therefore, in this sense, as in Mr. De Morgan's, is different on different occasions. (Peirce 1870, p. 366)

Thus, Peirce began with a class of individuals, namely a 'universe', which is a basic idea of model theory. Indeed, Peirce had a clear notion of a propositional function of many arguments on a domain D, with values 0 and 1. Hence his approach should be taken to be model-theoretic in that on the one hand the reasoning was about an interpretation under which a proposition is true, that is, a model of the proposition; and on the other the validity of inference was to be tested by truth with reference to that model. It follows that Peirce's work which was then developed by Schröder involved the main elements of the new research programme that Löwenheim initiated, namely the model theoretic programme which has become a very important part of modern logic.

In sum, in his 1885 paper, Peirce set up a language for first-order predicate logic and the rules for its use. Although his system was not as well-formed as Frege's, he developed a theory of quantification which is virtually equivalent to modern first-order predicate logic[2]. Peirce even distinguished between

---

[2] Peirce's development of the quantifiers should not give the misleading impression that his system was as well-formed as Frege's. For, unlike Frege, Peirce did not build a formal language since he did not introduce a formal notion of formula or of proof in his system.

first-order and second-order predicate logic[3] and treated the former by itself. Moreover, it was Peirce's notation and method that were employed by modern logicians, such as Löwenheim and Skolem in their model-theoretic approach to logic. Peirce was the first person to publish a paper that used explicitly the term 'quantifier': 'if the quantifying part, or Quantifier, contains $\sum_x \ldots$' (Peirce 1885a, p.183). As Church acknowledges it, 'the terms *quantifier* and *quantification* are Peirce's' (Church 1956, p. 288). Thus some years later than Frege, Peirce and Mitchell made their discovery of the quantifier without having ever taken knowledge of the work of Frege.

Then, these simple facts lead Putnam to say:

> Frege did 'discover' the quantifier in the sense of having the rightful claim to priority; but Peirce and his students discovered it in the effective sense. The fact is that until Russell appreciated what he had done, Frege was relatively obscure, and it was Peirce who seems to have been known to the entire world logical community. How many of the people who think that 'Frege invented logic' are aware of these facts? (Putnam 1992, p. 257)

Seemingly Dummett is not; otherwise he would not have been so outright when declaring that the use of quantifiers is due to Frege and altogether absent from Boole's logic. But, although the quantifiers are not present in Boole's work itself, it turns out that they were developed by his followers, that is, the members of the 'Algebraic School' who worked within the research programme established by Boole's logical calculus. Thus we have seen that Peirce was not satisfied with Boole's use of the symbol 'v' to express existential judgments (see subsection **3.6.2**). It is true that their dissatisfaction with this aspect of Boole's notation is in a way what led Peirce and his student O. H. Mitchell to their own discovery and treatment of quantifiers. Indeed, their discovery of the quantifiers provided a vast extension of the programme they had inherited from Boole for developing a calculus of deductive reasoning. Therefore, by using the discovery of the quantifiers as a criterion to separate modern logic from its precursors, Dummett fails to identify unequivocally where modern logic starts.

Yet, in his 1985 review of MacHale's book on Boole, Quine noted that 'logic became a substantial branch of mathematics only with the emergence of general quantification theory at the hands of Peirce and Frege' and thus dated modern logic from there. But Quine even conceded that:

---

[3] Peirce extended his system to what he called, 'second-intentional logic'. It contains a second family of variables, ranging over relations. Mathematical notions occurred in this second-intentional logic. For instance, he defined one-to-one correspondence employing second-order quantifiers (see Peirce 1885a, p. 188).

> The avenue from Boole through Peirce to the present is one of continuous development, and this, if anything, is the justification for dating modern logic from Boole; for there had been no comparable influence on Boole from his more primitive antecedents. (Houser 1997, p. 6)

Thus, even if Quine did not date modern logic directly from Boole, he did admit its origin from Peirce who squarely belonged to Boole's research programme. It follows that the continuous development of modern mathematical logic could not be precisely accounted for without including Boole's research programme. Indeed, after Leibniz's project of *calculus ratiocinator*, which formulated for the first time the idea of mathematical logic, it took almost two centuries until Boole came up with the great idea of introducing mathematics in logic. Thereafter, mathematical logic evolved continuously and the debate concerning the foundation of mathematics which was initiated by Frege at the beginning of the twentieth-century gave a boost to work in the area. This seems to be a 'fair-minded' account of the different 'moments', which stresses the historical importance of the different schools involved in the development of mathematical logic. It seems inconceivable, as Putnam sees it,

> That anyone could date the continuous effective development of modern mathematical logic from any point other than the appearance of Boole's two major logical works, the *Mathematical Analysis* and *The Laws of Thought*. (Putnam 1990, p. 255)

Boole carried out a formal system in that he built up an algebra of logic by means of what he called the 'general method', that is, the process by which combinations of interpretable symbols are carried out, according to well-determined rules of combination of the symbols in abstraction from their nature. This abstract formal use of symbols suggested to him that he should try to fit an algebraic structure to logical relations. Hence it is quite indefensible that Dummett denies in Boole's logical calculus the presence even in embryo of a formal system.

The 'general method' shows best the importance of Boole in the history of logic. The method is not simply a symbolical expression of ordinary language, which had been the chief concern of logic from Aristotle's time onwards. By means of his method, Boole treated logic for the first time mathematically. It allowed him to apply mathematical procedures to logic whilst preserving the basic independence of logic from mathematics. He brought forth a formal system, capable of various interpretations in different domains, from which stemmed the concept of 'Boolean algebra' which now plays a pivotal role in mathematical logic and computer design. Boole's formalism even enhanced metamathematical investigations, a distinctive feature of modern mathematical logic (see below sections **7.2** and **7.3**).

Furthermore, Boole's systematisation and generalisation of Aristotelian logic did not only encompass the traditional theory of the syllogism. It represented a significant transformation of the theory: before Boole, syllogistic theory was restricted to concern with premises which include exactly two terms. But when Boole implemented his computational procedure in syllogism, inferences which include any number of terms could be drawn. This allowed the analysis of all possible combinations, and thus the deduction of many conclusions consistent with the premises. Moreover, when Boole tacitly introduced the concept of the null class (see Boole 1847, p. 21), the Aristotelian view that all quantified propositions have existential import was abandoned. The modern view, due to Boole, is that existentially quantified propositions do have existential import and that universally quantified propositions do not. It should be also noted that Boole gave one clear conception of logical relations, though it pertained to de Morgan, who worked within his research programme, to develop a systematic formulation of a theory of relations, which had been left out in Aristotle's logical system.

Accordingly, Boole's work has a great logical importance in that it carried out a significant extension and improvement of the traditional theory of syllogism. Indeed, the generalisation and systematisation of Aristotelian logic is a significant advance in the development of mathematical logic. What is undeniable is that with Boole Aristotle's influence started to be loosened so as to allow a modern prospect of logic.

In addition, Dummett does not mention Boole's propositional calculus in which Boole had the idea of all possible distributions of truth-values. Thus, the method of truth-tables, which has been attributed to Post and Wittgenstein, can be traced back to Boole's calculus of hypothetical propositions. It follows that Boole went well beyond the simple creation of a general theory of classes which encompasses the traditional theory of syllogism. He built up a calculus of propositions, presenting its earliest systematic formulation, at least since the Stoics. This reinforces the importance of Boole in the history of logic. As Kneale puts it,

> The chief novelty in Boole's system is his theory of elective functions and their development, or, as we should now say, his theory of truth-functions and their expressions in disjunctive normal form. Philo of Megara discussed particular elective functions and explained how they could be developed, but Boole should have the credit of being the first to treat these two topics in general fashion. (Kneale 1962, p. 420)

It turns out that Boole played an important role in the halfway period between the Aristotelian and the Fregean research programmes. He carried out the leading principle of the research programme of the 'Algebraic School', which was then enriched and systematically developed by Peirce and Schröder who opened up the model-theoretic approach to logic.

Now, an appraisal that does justice to Boole's work was given by McColl:

> The first person to show that symbolical reasoning might also be employed with advantage in the investigation of matters usually considered altogether beyond the sphere of mathematics was the late Professor Boole. This he did first in his *Mathematical Analysis of Logic*, and afterwards more fully in his celebrated *The Laws of Thought*, published in 1854. These works excited much admiration in the mathematical world, and, it may be added, caused no small trepidation among logicians, who saw their hitherto inviolate territory now for the first time invaded by a foreign power, and with weapons which they had too much reason to dread. (McColl 1880, p. 46)

7.2 Boole's Anticipation of Metamathematics

Metamathematics was furthered by those working within Hilbert, Löwenheim and Tarski's research programmes, in which the modern distinction between the syntactic (proof-theoretic) and the semantic (model-theoretic) approach of the notion of logical consequence is explicitly stated. These works on metamathematics led to the highest point in the development of mathematical logic with the achievements of Gödel. Of particular interest was the publication of his incompleteness theorems in 1931, which rendered very doubtful the Frege-Russell and Whitehead reductionist programme as well as the Hilbert programme of a 'complete' mathematics.

I shall argue that the very possibility of metamathematical investigations emerged with Boole, for such inquiry supposes the existence of formal systems capable of various interpretations that can be subjected to mathematical study.

7.2.1 Boole's Logic as Formalism

Boole wanted to extend mathematics by establishing an abstract view of mathematical operations without regard to the objects of these operations. He claimed 'a place among the acknowledged forms of Mathematical Analysis' (Boole 1847, p. 4) for the calculus of logic. But, by doing so, he did not

Metamathematics: A Return to Boole's Research Programme   233

merely want to include logic in traditional mathematics. Rather, he thought of a new mathematics which is expressed as follows: 'it is not the essence of mathematics to be conversant with the ideas of number and quantity' (Boole 1854 p. 12). Thus Boole set up algebraic systems of related operations, and articulated a common abstract structure which led him to formulate a set of axioms that are satisfied by each of the systems. Amongst these systems was logic which Boole constructed as a formal process of reasoning depending only upon the laws of the symbols, and not upon the nature of their interpretation.

Boole's formalist treatment of logic led on to Hilbert's metamathematics. In effect, Hilbert was concerned with notions about syntax and proof and aimed to shed light on formal languages and deductive systems. Unlike Frege, Russell and Whitehead, he did not believe that arithmetic could be reduced to logic. Rather, he argued that both should be integrated in an axiomatic system of which the postulates are then proved consistent, complete, and independent. He called the project of providing such proofs metamathematics. In order to carry out this project, Hilbert needed a symbolic language. For him the purpose of this symbolic language in mathematical logic is to achieve in logic what it has achieved in mathematics, namely, an exact scientific treatment of its subject-matter. In their book, *Principles of Mathematical Logic*, Hilbert and Ackermann wrote:

> The logical relations which hold with regard to judgements, concepts, etc., are represented by formulas whose interpretation is free from the ambiguities so common in ordinary language. The transition from statements to their logical consequences, as occurs in the drawing of conclusions, is analysed into its primitive elements, and appears as a formal transformation of the initial formulas in accordance with certain rules, similar to the rules of algebra. (Hilbert and Ackermann 1928, p.1)

Thus, like Boole, Hilbert conceived logic as a formal process which proceeds by the combination of pure symbols, subject to certain rules and in abstraction of their meaning. Hence logic is a play of symbols, according to certain rules. It may be even said that with Hilbert logic is a logic of symbols. As he himself said, 'what we consider is the concrete signs themselves, whose shape... is immediately clear and recognizable' (Hilbert 1925, 376). Here is the core of formalism: logic and mathematics, in particular, are about symbols.

Hilbert was quite conscious of what the algebra of logic, with its emphasis on algorithms, could mean for all mathematics. For once logical formalism is established one can expect that a systematic computational treatment of logical formulas is possible, which would somewhat correspond to the theory of equations in algebra. As he said,

> We may proceed in the logical calculus just as we do in algebra, where we write formulas with letters which mean that for any arbitrary numerals that we substitute for the variables the resulting numerical equation is true. (Hilbert and Ackermann 1928, p. 58)

Thus Hilbert's conception of logical calculus is akin to Boole's idea of using an algebraic system as an algorithm, handling the symbols purely mechanically without any dependence on meanings. Such an idea is now called 'disinterpretation'. Dummett finds the same analogy between Boole and Hilbert in their manner of construing a mathematical theory. As he points out,

> For Hilbert, a definite individual content... may legitimately be ascribed only to a very narrow range of statements of elementary number theory... All other statements of mathematics are devoid of such a content... The other mathematical statements are not... devoid of significance: but their significance lies wholly in the rôle which they play within the mathematical theories to which they belong, and which are themselves significant precisely because they enable us to establish the correctness of finitistic statements. Boole likewise distinguished, amongst the formulas of his logical calculus, those which were interpretable from those which were uninterpretable: a deduction might lead from some interpretable formulas as premisses, via uninterpretable formulas as intermediate steps, to a conclusion which was once more interpretable. (Dummett 1978, p. 219)

Hilbert believed that a mathematical theory is to be constructed strictly formally. For him, any mathematical theory can be firstly completely formalised and then subjected to a mathematical analysis. This formalisation is to be done in an axiomatic manner. He thought of the axiomatic method as the instrument suited to the human mind and indispensable for all scientific thinking. But the language which he used in a mathematical theory is different from the language which he used in order to analysis that theory. Hilbert considered the mathematical language as a separate item and studied it as mathematical language itself.

Boole's treatment of algebra as an abstract structure may be taken as a heuristic preliminary for Hilbert's construction of a formal mathematical theory and a language about it. Indeed, like Hilbert who advocated the axiomatic method, Boole believed that

> All sciences consist of general truths, but of those truths some only are primary and fundamental, others are secondary and derived...

> And it is so also in the purely mathematical sciences. An almost boundless diversity of theorems, which are known, and an infinite possibility of others, as yet unknown, rest together upon the foundation of a few simple axioms; and yet these are all general truths. (Boole 1854 p. 5)

Boole did not explicitly define the syntactic notion of logical consequence: but it is implicit in his logical calculus. For there is the formal system, with its symbols and rules of formation, axioms and laws of derivation. Thus, Boole would have had the material necessary to formulate the notion of completeness. Moreover, although he did not use completeness in the modern technical sense, he toyed with the notion when asking

> Whether in any science, viewed either as a system of truth or as the foundation of a practical art, there can properly be any other test of the completeness and the fundamental character of its laws. (Boole 1854 p. 5)

Boole went beyond the mere analytical processes and investigated the condition of a perfect method in the construction of the science of reasoning. He inquired about what is best as respects not only the mode or form of deduction, but also the system of premises from which the deduction is to be made (Boole 1854, p. 150). This led him to assess his method as possessing 'a theoretical unity and completeness which render it deserving of regard'. (Boole 1854, p. 157)

However, the existence in Boole's logical system of material necessary to formulate the notion of completeness was known to Peirce. In effect, in his 1885 paper, Peirce exhibited the pattern of reasoning that the truth-tables tabulate and suggested the general notion of validity in the propositional calculus as follows:

> To find whether a formula is necessary true substitute f and v for the letters and see whether it can be supposed false by any such assignment of values. (Peirce 1885a, p. 175)

For Peirce **v** and **f** are two constant values. If the value of a proposition is **v**, the proposition is true, and if it is **f** the proposition is false. Boole chose **v** = 1, **f** = 0. As an example, Peirce took the formula

$$(x < y) < ((y < z) < (x < z)).$$

For convenience of notation, I use here the sign '<' instead of Peirce's inclusion (illation) sign. Thus to make the formula false we must take

$$(x < y) = \mathbf{v}$$
$$((y < z) < (x < z)) = \mathbf{f}.$$

The last gives

$$(y < z) = \mathbf{v}$$
$$(x < z) = \mathbf{f}$$
$$x = \mathbf{v}$$
$$z = \mathbf{f}.$$

Substituting these values in

$$(x < y) = \mathbf{v}$$
$$(y < z) = \mathbf{v}$$

we have

$$(\mathbf{v} < y) = \mathbf{v}$$
$$(y < \mathbf{f}) = \mathbf{v}$$

which cannot be satisfied together.

This notion of validity makes possible a notion of completeness of formal systems of the propositional calculus: a system is complete if every valid formula is a theorem. According to Hilbert and Ackermann, the familiar account of the completeness of an axiom system is that a system is complete if all the valid formulas of a certain domain which is characterised by content can be proved from the set of axioms[4] (Hilbert and Ackermann 1928, p. 42). Hilbert and Ackermann then gave an explicit proof of the completeness of the propositional calculus in this sense (see p.43). If now Boolean algebra with two elements can be regarded as a propositional calculus, then a completeness proof can be extracted from developed normal forms which were due to Boole (see subsections **3.2.1** and **3.5.1**). Moreover Hilbert and Ackermann showed that the decision problem for the sentential calculus is also solved by means of a

---

[4]Hilbert and Ackermann also defined the completeness of an axiom system in an unfamiliar sense, that is, 'an axiom system is termed complete only if a contradiction always arises when there is added to the axioms a formula not previously provable from them' (Hilbert and Ackermann 1928, p. 42). They gave an explicit proof of the completeness of the propositional calculus in this sense (see p. 43).

Metamathematics: A Return to Boole's Research Programme 237

technique using Boolean developed normal forms (see Hilbert and Ackermann 1928, pp.17–18).

In addition, although Boole did not construct one, he suggested an axiomatic method as the form under which all systems of truth should be presented (1854, p. 5). This would then imply implicitly the metamathematical investigations into the axioms adopted in this system. Hence, when Boole's followers, such as Sheffer and Huntington, built up their sets of independent postulates for Boolean algebra, they were explicitly concerned with the proof of the consistency and the independence of the postulates of the axiomatic system of Boole's logical calculus (see section **4.3**).

Boole conceived his algebra as a logical structure which admits of a variety of interpretations, and as these are independent of each other, it follows that the structure is independent of any of them. The concepts and operations of Boole's algebra can be interpreted as applying to classes, propositions, electronic circuits[5], probabilities and several other mathematical domains. For instance, Huntington interprets them as applying to geometry, that is, to overlapping plane compartments of space (see section **4.2**). On the other hand, unlike Boole, Frege thought of logic as a universal language embracing a universe of all conceptions, and thus it is impossible to 'step outside' of such a language so as to raise metamathematical questions. Moreover, in Frege's language of predicate logic 'every logical formula has a fixed meaning; there is no question of reinterpreting any sign' (Goldfarb 1979, p. 354).

It follows that Boole's formal system, capable of various interpretations in different calculi, was likely to have suggested the actual work on metamathematics inquiring into the validity of well-formed formulas in these different calculi. Boole thus opened a new avenue for mathematicians to consider the interpretations of their languages and metasystematic questions about their systems.

7.2.2 Boole's Logic as Semantics

Boole gave various different interpretations of his logical calculus, interpreting the symbols either as classes, or as propositions, or even as periods of time. He suggested correspondences between these interpretations and thereby could be regarded as developing an early formal semantic theory. He stressed that it is necessary that each symbol of his logical system should possess, within the limits of the same discourse or process of reasoning, a fixed interpretation.

---

[5] In 1938, C. E. Shannon applied Boolean algebra to the study of switch and relay circuits in electrical communication engineering.

But what does the symbols stand for in the processes of reasoning? For Boole, the general answer to this question is,

> That in the processes of reasoning, signs stand in the place and fulfil the office of the conceptions and operations of the mind; but that as those conceptions and operations represent things, and the connexions and relations of things, so signs represent things with their connexions and relations; and lastly, that as signs stand in the place of the conceptions and operations of the mind, they are subject to the laws of those conceptions and operations. (Boole 1854, p. 26)

Thus Boole distinguished between the syntactic relations among signs themselves and the semantic relations between signs and things. He pointed out that signs are subject to fixed laws of combination depending upon the nature of its interpretation. By referring to relations between signs and other things in the world or their meanings, Boole specified the semantical aspect of his logical system.

Moreover, without going as far as Löwenheim and Tarski, Boole engaged in metatheoretical argumentation by raising semantic problems when he inquired about the conditions of valid reasoning by the aid of symbols. Thus he stated semantical rules in the metalanguage that concerns the meaning of expressions in the object language. These rules require that (1) a fixed interpretation is assigned to the symbols employed in the expression of the data; and that the laws of the combination of those symbols are correctly determined from that interpretation; that (2) the formal process of demonstration is conducted according to the laws without regard to the question of the interpretability of the results obtained; and that (3) the final result is interpretable in accordance with that system of interpretation which has been employed in the expression of the data. Boole observed that the necessity that the fixed interpretation of the symbols should be in such a form as to admit of that interpretation being applied is founded on the principle that

> The use of symbols is a means towards an end, that end being the knowledge of some intelligible fact or truth. And that this end may be attained, the final result which expresses the symbolical conclusion must be in an interpretable form. (Boole 1854, p. 68)

This semantical principle which is based on the notion of interpretation and Boole's term of 'universe of discourse' suggested what is now known as model theory. For an interpretation under which the final result is true is said to be a model of the system. Boole's logical system can be interpreted as a theory of propositional functions of any number of variables with values 0, 1 on a domain D which is a basic idea of model theoretic approach to logic. Furthermore, this interpretation can exhibit a technique i.e. a decision procedure for testing validity which involves the concepts of truth and falsity. Indeed, Boole's developed normal forms may be construed as the truth-table representation of a logical function, inasmuch as it pictures both those constituents corresponding to truth-possibilities, which the formula matches, as well as those which it does not (see subsection **3.5.1**). This test involves picturing circumstances that would make compound propositions true or false. Such circumstances may be thought of as 'models' of the possible truth-conditions, and the test that involves such considerations are therefore called model-theoretic. Hence it appears that Boole's approach to logic was in this respect model theoretic.

Model theory is the study of interpretations and their properties. It was initiated by Löwenheim and then developed by Tarski who was concerned with the methodology of deductive systems. Tarski analysed central semantic notions, such as logical consequence, satisfiability, and decidability, which direct the course of various ongoing research programmes.[6] As will be shown below, it turns out that Löwenheim's theorem of 1915, which gave rise to the model-theoretic treatment to logic, required a formal system capable of various interpretations in different domains of discourse, such as Boole's logical calculus.

From all this it can be concluded that, by the simple fact that his abstract algebraic structure raised syntactic and semantic problems, Boole should earn the right to be considered as the grandfather of metamathematics.

7.3 Löwenheim's Revival of Boole's Research Programme

Metamathematical investigations of the model-theoretic kind first appeared in Löwenheim's paper 'On Possibilities in the Calculus of Relatives' (1915) in which he was concerned with Peirce and Schröder's work in Boolean algebra, and not, as it might have been expected, with the logic of Frege or Russell and Whitehead's *Principia Mathematica*. The historical importance of the paper is stressed by Van Heijenoort who concludes his 'Logic as Calculus and Logic as Language', by noting that 'after *Begriffsschrift* (1879), Löwenheim's paper (1915), and Chapter 5 of Herbrand's thesis (1929) are the three cornerstones of modern logic' (Van Heijenoort 1967b, pp. 328–339).

In effect, Löwenheim's paper appeared as a pioneer work in mathematical logic. In this paper, which undoubtedly belongs to the algebraic trend in logic,

---

[6]See Tarski 1956.

Löwenheim revived Boole's research programme. He was mainly concerned with the validity, in different domains, of formulas of first-order predicate calculus with identity and the decision problems. He denoted the universe of discourse by 1^1, that is, a basic domain of individuals consisting of at least one element. It may be finite or infinite and if infinite, countable or not. He used Schröder's notation, the logic of relatives and the notation of the universal and existential quantifiers introduced by Peirce, and the Boolean expansion, in terms of the distinguished classes, as a disjunctive normal form. Also, Löwenheim began his paper by treating a calculus of relatives in which the quantifiers and the relative coefficients are eliminated. As Heijenoort notes, 'the calculus thus obtained contains variables and constants for binary relations, and the operations are the Boolean operations, together with the converse, the relative product, and the relative sum' (Van Heijenoort 1967a, p. 229).

The results of the content of Löwenheim's paper regarding the first-order predicate calculus with identity are recapitulated by Church as follows: a solution of the decision problem for validity in the case that only individual predicate variables appear; a reduction of the general case of the decision problem for validity to that in which only binary predicate variables appear; recognition of the existence of well-formed formulas that are valid in every finite domain but not valid in an infinite domain, and a demonstration that no well-formed formula containing only individual predicate variables can have this property; finally, a proof of the metatheorem known as Löwenheim's theorem (Church 1956, p. 293).

This famous Löwenheim's theorem says that

> If the domain is at least denumerably infinite, it is no longer the case that a first-order fleeing equation is satisfied for arbitrary values of the relative coefficients. (Löwenheim 1915, p. 235)

Löwenheim defined a fleeing equation as 'an equation that is not satisfied in every 1^1 but is satisfied in every finite 1^1' (Löwenheim 1915, p. 233). Church rewords the theorem as follows: 'if a well-formed formula is valid in an enumerably infinite domain, it is valid in every non-empty domain' (Church 1956, p. 238). In 1920 the theorem was restated and reproved by Skolem, to yield the famous Löwenheim-Skolem theorem (see Skolem 1920, p. 256).

It is worth noting that the first application that Löwenheim made of his theorem is to the Boolean calculus of classes. As he said, 'all questions concerning the dependence or independence of Schröder, Müller, or Huntington's class axioms are decidable (if at all) already in a denumerable domain' (Löwenheim 1915, p. 240). Furthermore the theorem is entirely proved in Peirce's notation which makes Putnam say that 'first-order logic and its metamathematical study would have existed without Frege' (Putnam 1992, p. 258).

Metamathematics: A Return to Boole's Research Programme

Löwenheim distinguished quantification over individuals from quantification over relatives, and showed that his theorem cannot be proved for higher-order logic. Thus for the first time a distinction is made between first-order and higher-order logics. Indeed, Löwenheim carefully delineated the class of first-order expressions which would later be known as the first-order part of a logical system. But, as Goldfarb notes in his paper on the nature of the quantifier, 'Löwenheim's interest in the first-order fragment of the relative calculus seems motivated by purely algebraic, rather than foundational considerations' (Goldfarb 1979, p. 355). Nonetheless Löwenheim claimed that

> Every theorem of mathematics, or any calculus that can be invented, can be written as a relative equation; the mathematical theorem then stands or falls according as the equation is satisfied or not. This transformation of arbitrary mathematical theorems into relative equations can be carried out, I believe, by anyone who knows the work of Whitehead and Russell. (Löwenheim 1915, p. 246)

Löwenheim called 'relative equation' a first-order equation. Thus he thought that all mathematical judgements can be represented by first-order expressions in the calculus of relations. But the possibility of this remark within the framework of Löwenheim appears to be doubtful. Does the relative calculus, which contains variables only over individuals in a domain and over sets and relations of individuals, have the enormous expressive capacity required to capture mathematical inferences involving high-order notions? Moreover, according to Goldfarb, Löwenheim did not have a full sense of the role of the object language in formalisation of mathematics, and the absence of formal inference rules prevents the use of the relative calculus for axiomatisation, in the sense of formal systems (Goldfarb 1979, p. 355). In effect, as Goldfarb sees it, 'Löwenheim lacks a general notion... of a formalized mathematical theory, which could encompass systems with infinitely many axioms'. However, 'without such a notion, there can be little question of providing first-order encodings of mathematics' (Goldfarb 1979, p. 355).

The considerations of the validity of formulas in different domains which underlay Löwenheim's theorem were rendered possible by Boole's logical calculus, in which the universe of discourse denoted by 1 could vary from context to context, and thus contains only what we concur to consider at a certain time, in a certain context. As Boole said,

> ...whatever may be the extent of the field within which all the objects of our discourse are found, that field may properly be termed the universe of discourse. (Boole 1854, p. 42)

Since Boole treated his logical system as a formal calculus capable of various interpretations with different domains of discourse, hence arises the possibility of demonstrating the validity of well-formed formulas in different domains of discourse.

Thereafter, Boole's followers, such as Peirce and Schröder, carried out the calculus of relatives in which the quantifiers occur as certain possibly infinite sums and products over individuals or relations. They wanted to apply logical formulas to different universes of discourse and to treat first-order predicate calculus by itself. Goldfarb remarks that they investigated the following question:

> Given an equation between two expressions of the calculus, can that equation be satisfied in various domains-that is, are there relations on the domain that make the equation true? (Goldfarb 1979, p. 354)

Such an inquiry is very much like the modern notion of satisfiability of logical formulas. As Heijenoort remarks, 'behind the Frege-Russell trend in logic, Löwenheim renews contact with Boole and Schröder, while making important contributions of his own to logic' (Van Heijenoort 1967b, p. 228).

For Heijenoort and Goldfarb, metamathematical considerations are ruled out in the logicist programme, because Frege set up his logical system as a universal language in which the quantifiers binding individual variables range over all objects. Indeed Frege treated first-order predicate calculus as a part of an ideal language with a fixed universe of discourse consisting of all there is, namely, 'all objects'. The universe of discourse of this *lingua characterica* is the universe. As Goldfarb puts it,

> For Frege we may be speaking of all objects or all functions; for Russell of all individuals or all propositional functions of some particular order. The ranges of the quantifiers are fixed in advance once and for all. The universe of the discourse is always the universe, appropriately striated. (Goldfarb 1979, p. 352)

Thus this *lingua* is completely universal, and therefore there is no interesting perspective 'outside' Frege's logical system from which to study it. Wittgenstein, who was vehemently opposed to any metatheoretical investigations opened, his Notebooks (1914) as follows:

> Logic must take care of itself. If syntactical rules for functions can be set up *at all* for functions, then the whole theory of things, properties, etc., is superfluous. It is also all too obvious that this

theory isn't what is in question either in the *Grundgesetze*, or in *Principia Mathematica*. (Wittgenstein 1914, p. 2)

However, although metamathematical considerations seemed to be more or less alien to Frege's research programme, Heijenoort and Goldfarb's conviction that semantic notions are unknown in *Principia Mathematica* as they are in Frege's work should be moderated. Indeed, in the introduction of the *Tractacus Logico-Philosophicus*, Russell demurred from Wittgenstein's argument against the possibility of metamathematics. He wrote:

> These difficulties suggest to my mind some such possibility as this: that every language has, as Mr Wittgenstein says, a structure concerning which, in the language, nothing can be said, but that there may be another language dealing with the structure of the first language, and having itself a new structure, and that to this hierarchy of languages there may be no limit. (Wittgenstein 1913, p. xxii)

In *The Basic Laws of Arithmetic*, Frege too tried to show that proper names, names of first-level functions, and every proposition of *Begriffschrift* have a denotation (Frege 1893, pp. 83–89). This seemed to hint at the possibility of a semantic theory. As Dummett held it,

> Frege's notion of reference is best approached via the semantics which he introduced for formulas of the language of predicate logic. An interpretation of such formula...is obtained by assigning entities of suitable kinds to the primitive non-logical constants occurring in the formulas...this procedure is exactly the same as the modern semantic treatment of the language of predicate logic. (Dummett 1981b, pp. 89–90)

Dummett could therefore claim that Frege had had within his grasp the concepts necessary to frame the notion of the completeness of a formalisation of logic, as well as its soundness (Dummett 1981b, p. 82). Moreover, once again, it is regrettable that the correspondence between Frege and Löwenheim was lost in the Second World War. For it is said that Löwenheim had written to Frege in 1908 to argue for the possibility of a purely formal arithmetic on the basis of considerations from the second volume of *The Basic Laws of Arithmetic*, and he even convinced him of such a possibility.

Nonetheless, it remains true that metamathematical considerations of the model-theoretic kind arose from Boole's research programme and not from Frege's. But it should be stressed as well that they have reconnected the two research programmes which had been separated and classified respectively under the headings of 'logic as calculus' and 'logic as language'. Indeed the development of mathematical logic in the first few decades of the last century is a blending of the two research programmes. In this sense, Löwenheim's theorem, which is a revival of Boole's research programme, is a metamathematical study of the first-order predicate calculus which is exhibited in Frege's *Begriffsschrift*.

Heijenoort writes that 'with Löwenheim's paper we have a sharp break with the Frege-Russell approach to the foundations of logic and a return to, or at least a connection with, pre-Fregean or non-Fregean logic' (Van Heijenoort 1967b, p. 328). But the break is not as sharp as Heijenoort suggests, for through Löwenheim the Boolean and Fregean research programmes are somehow connected. Moreover, the gulf between 'logic as language' and 'logic as calculus' does not seem to be unbridgeable. Indeed, semantic notions were not foreign to logicians who considered language as fully interpreted, universal, and referring only to one world. There was a great deal of interplay between logicians on one hand and metamathematicians and algebraists on the other. In truth, modern logic is a descendant of the interaction of the logicist and the algebraic research programmes.

7.4 Back to Gillies' 'The Fregean Revolution in Logic'

In the introduction of his paper 'The Fregean Revolution in Logic', Gillies considers the beginning of the Fregean revolution with the publication in 1879 of the *Begriffsschrift*. Although he recognises that opinions may differ as to when the revolutionary period ended, he sees the publication of Gödel's incompleteness theorem in 1931 as forming a natural terminus (Gillies 1992, p. 265). Gillies' general conclusion is this:

> Frege's *Begriffsschrift* undoubtedly contributed a number of central conceptions to the new paradigm for logic, but the *Begriffsschrift* did not become the canonical text of the new logic. The semantic side of logic was developed independently of Frege, while the ideas of the *Begriffsschrift* itself were taken up and developed by logicians such as Peano, Russell, Hilbert, Carnap, and Church. In the process, Frege's original system was modified in many respects. (Gillies 1992, pp. 276–277)

Thus, Gillies seems to consider the subsequent development of mathematical logic after Frege and before Gödel as a part of the Fregean Revolution. It has already been argued that Gillies' application of Kuhn's concept of 'extraordinary science' to logic leads to an oversimplification of the development of logic consisting of reducing the whole logical enterprise into Aristotle, Frege, and Gödel. Thus the concept of research programme has been preferred over that of 'revolution' for discussing the development of logic (see section **1.1**). But even if it is assumed that the Kuhnian analysis of revolutions which Gillies basically follows is appropriate for studying the history of logic, it would then follow that the emergence of the semantic side of logic should be considered as a new paradigm. For it was an achievement which showed the two characteristics to which Kuhn refers as a definition of paradigms, that is, a sufficiently 'unprecedented' and 'open-ended' achievement which attracted a community of mathematicians and logicians away from the predominated logicist programme (Kuhn 1962, p. 10).

Gillies himself admits that the semantic side of logic was developed independently of Frege. But, if granted that the concept of research programme is the core unit for studying the development logic, then the model-theoretic approach to logic should be regarded as a new research programme distinct from the Fregean research programme in that it would be a change from a Fregean programme, whose core was the syntactic axiomatisation of logic, to a programme in which semantic considerations prevail. This research programme would begin with the publication of Löwenheim's paper of 1915, which would perform a rupture with Frege's programme. In effect, before Löwenheim's paper there was no semantical analysis concerned with the connection between an axiomatic-deductive system and its interpretations or models. The characterisation of the conditions under which a formula is true in a model was an unprecedented study in logic.

Löwenheim outlined the research programme within which Skolem and Herbrand carried out their work. Furthermore, the method used by Gödel to show, for the first time, the links between the semantic concepts of validity and satisfiability of a formula and the syntactic concept of provability implies Löwenheim's theorem. Following subsequent work of Tarski, the model-theoretic approach to logic at present constitutes an extremely active open-ended area of research. For instance, in their book *Model Theory*, Chang and Keisler (1973) include some of the more recent results which are the Keisler-Shelah isomorphism theorem, the Morley categoricity theorem, the work of Ax-Kochen and Ershov in field theory, and the results of Rowbottom, Gaifman and Silver on large cardinals and the constructible universe. These results which are stimulating present and future model-theoretic research cannot be simply considered as a part of Frege's original system. Rather, they belong to a new research programme.

However, it should be noted that the advance of the semantic side of logic would complement without completely discarding the Fregean research programme. In this connection, Gillies borrows from Kuhn the analogy between political revolutions and scientific revolutions for the subsequent application to logic. Although his general view is that revolutions in mathematics are of the 'Franco-British' type in which 'something' is not 'overthrown and irrevocably discarded' but simply loses its former importance, he compares the Fregean revolution in logic to the 'Russian revolution' in which the 'something' is 'overthrown and irrevocably discarded' (Gillies 1992, p. 269). Accordingly, he would consider that there is an absolute and radical rupture of continuity in the development of logic. This would explain why he seems to minimise the importance of the research programme of Boole and its further refinements.

But logical research programmes do not seem to involve such an absolute and outright breach of continuity. Rather, there is a continuous development of mathematical logic which is made up of overlapping research programmes that share the so-called 'common logic'. Whether one considers the Fregean revolution or the Gödelian revolution, in each case the continuities in conceptual and theoretical structure exist. A revolutionary period in logic is not a period in which a 'routine' is overthrown and replaced by another. When Gillies says that there is a Fregean revolution, I agree. Because the concept of 'revolution' can be used to express the historical importance of Frege's great achievement, and the analogy to political revolution can be regarded as a means of expression for the historian. But I disagree when Gillies applies Kuhn's concept of 'normal scientists' to all those logicians who did not initiate, by a '*Gestalt* switch', an extraordinary revolution. In effect, following Kuhn's view, Gillies presumably thinks that there are two different historical periods to be distinguished in the history of logic: (1) normal periods in which there is an established 'routine' of puzzle solving; (2) extraordinary period consisting of a crisis followed by a revolution. But this classification does not fit the history of logic. For instance, where would Gillies classify the splendid achievement of the model-theoretic approach to logic? He has to regard it as part of the new established 'routine' of the Fregean paradigm, so that the development of model theory turns out to be just 'normal logic'.

Against this I put forward an alternative between Kuhn's dualistic scheme. This alternative is built upon the concept of research programme without venturing into Lakatos' specific ideas about the concept. This concept of research programme can play a methodological role in the historiography of logic by guiding and helping research in this area. There is much evidence to suggest, indeed, that the history of logic is marked by overlapping and interpenetrating research programmes whose involvements sustain the development of logic, and this would make the focus on 'revolution' misleading. For, once it is recognised that the non revolutionary periods are not characterised by the

Metamathematics: A Return to Boole's Research Programme 247

application of a 'routine', but the modification and extension of old research programmes, the historical interest should become the treatment of different research programmes in order to find out what they have in common, what is not puzzle-solving about them and why this is so.

In this dissertation, I have tried to show how Boole's work initiated a research programme through which logicians used algebra to develop logic well beyond Aristotelian syllogistic. I also tried to show that the Boolean research programme, within which a theory of quantification and relational predicates were developed, is not completely overthrown either by Frege's work or the work of later logicians who used Frege's revolutionary insights. Instead the Boolean research programme led to the development of the model-theoretic approach to logic which is today recognised as complementing the proof-theoretic approach that is rightly regarded as Fregean in its foundation.

7.5 An Overall Portrait of Boole and Frege

Although Boole and Frege's research programmes were developed independently and were historically distinct, they were not antagonistic. Rather, Frege's research programme extended over and covered a part of Boole's. Thus Frege's logical system could be regarded as an integrated whole in which his logic, beginning from different basics, embodies Boole's as a fragment of his first-order predicate calculus. I shall here stress again some aspects of the close relationship between Boole and Frege.

For Boole, the laws of thought do not depend upon our will, and the forms of the science, of which they constitute the basis, are independent of individual choice. They have nothing to do with mental processes except to be conveyed via this path. He thus departed from psychologism by regarding the laws of logic as objective, and prescribing the character of its mathematical form and forbidding any other form of manifestation that is not scientific. He presented logic as a system of principles which allow for valid inference in all kinds of discipline. Hence Boole regarded logical laws as normative and universal. Such a conception of logic dovetails with Frege's principle which affirms the separation of the logical from the external world and the realm of subjective ideas.

Boole regarded logic as concerned with the study not of content, but of the form and structure of propositions. As Bochenski sees it, 'what is quite new in Boole, and in contrast to the whole tradition, Leibniz included, is that he did not think of logic as an abstraction from actual process, as all previous logicians had done, but as a formal construction for which an interpretation is sought only subsequently' (Bochenski 1970, 279). Kneale also notes that in expressing truths of logic in symbolism derived from mathematics Boole was already interested in the kind of formal study which is required for the solution of questions about the foundations of mathematics (Kneale 1947, p.158). It is indeed the construction of a formal system of logic that provides the connection between Boole and Frege. The latter set up a formal process of analysing logical relations when working out the problem of the foundations of mathematics. They both held that logic was concerned with the form of reasoning independent of the language within which it was expressed.

Boole employed his symbolic notation to formalise propositional calculus. In the *Mathematical Analysis of Logic*, he even developed a truth-functional analysis of propositional logic. His 'cases or conjunctures of circumstances' hinted at the notion of 'truth-value'. As Sluga admits, 'it is even plausible to assume that Frege modelled his propositional logic on that of Boole. Both assume naturally that propositions are either true or false and that they cannot be both at once. Both consider the truth and falsity of a complex proposition to be determined by the truth and falsity of the simple propositional components. For both, in short, propositional logic is two-valued and truth-functional' (Sluga 1980, p. 80). However, Frege extended the propositional calculus to a more general theory, namely, the predicate calculus of which Boole's propositional calculus constitutes a part.

The overall portrait of Frege and Boole that has now to be drawn was portrayed by Frege himself in the essay, 'On Mr. Peano's Conceptual Notation and My Own,' where he stressed that from another point of view his logical system is closer to Boole's than Peano's. He wrote,

> We can recognise a closer affinity between Boolean logic and my conceptual notation, in as much as the main emphasis is on inference, which is not stressed so much in the Peano logical calculus. (Frege 1897b p. 242)

In correspondence addressed to Hilbert on the 1st of October 1895, Frege held the same view by employing a botanic metaphor, the process of lignification, so as to picture the close relation between his concept-script and Boole's logical calculus:

> Where a tree lives and grows it must be soft and succulent. But if what was succulent did not in time turn into wood, the tree could not reach a significant height. On the other hand, when all that was green has turned into wood, the tree ceases to grow. (Frege 1895c p. 33)

Metamathematics: A Return to Boole's Research Programme 249

If I imagine now Frege's *Begriffsschrift* as a cherry tree blooming, then Boole's *The Laws of Thought* is the hard fibrous lignified substance under the bark of such a tree.

References

This list contains some books which have been used but not referred to elsewhere.

Aristotle (350B.C.E), *Prior Analytics*, translated by A.J. Jenkinson (accessible at http://classics.mit.edu/Aristotle/prior.1.i.html).

Aristotle (1949), *The Complete Works of Aristotle*, edited by Jonathan Barnes, Princeton, N. J., Princeton U. P., vol. 1 and 2.

Arnauld, A. & Nicole, P. (1662), *The Art of Thinking*, The Bobbs-Merrill Company 1964.

Ashworth, E. J. (1978), 'Multiple Quantification and the Use of Special Quantifiers in Early Sixteenth Century Logic', in *Notre Dame Journal of Formal Logic*, vol. 29, 4, pp. 599-613.

Baker, G. P & Hacker (1984), *Frege: Logical Excavations*, Oxford University Press, New York

Beaney, Michael (1997), *The Frege Reader*, Blackwell.

Bell, E. T. (1953), *Men of Mathematics*, Penguin Books.

Bentham, George (1827), *Outline of a New System of Logic*, London.

Beth, E. W. (1965) *Mathematical Thought*, Dordrecht-Holland.

Bochenski, I. B. (1970), *A History of Logic*, Chelsea Publishing Company New York.

Boehner, P. (1952), *Medieval Logic: An Outline of its Development From 1250 to c. 1400*, Manchester University Press.

Boole, George (1841a), 'On Certain Theorems in The Calculus of Variations', in *The Cambridge Mathematical Journal*, vol.2, pp. 97-102.

Boole, George (1841b), 'Researches On The Theory of Analytical Transformations', in *The Cambridge Mathematical Journal*, vol.2, pp. 64-73.

Boole, George (1841c), 'On The Integration of Linear Differential Equations With Constant Coefficients', in *The Cambridge Mathematical Journal*, vol.2, pp. 114-119.

Boole, George (1844), 'On a General Method in Analysis', *Philosophical Transactions of the Royal Society of London*, 34, pp. 225-282.

Boole, George (1847), *The Mathematical Analysis of Logic*, Cambridge.

Boole, George (1848a), 'The Nature of Logic and the philosophy of Mathematics', in *Selected Manuscripts on Logic and its Philosophy*, edited by Ivor Grattan-Guinness and Gérard Bornet, Birkhäuser Verlag, Basel, Boston, Berlin 1997, pp.1-12.

Boole, George (1848b), 'The Calculus of Logic', in *Studies in Logic and Probabilities*, edited by Rhees, London 1952, pp. 124-140.

Boole, George (1849), 'Elementary Treatise on Logic not Mathematical', in *Selected Manuscripts on Logic and its Philosophy*, edited by Ivor Grattan-Guinness and Gérard Bornet, Birkhäuser Verlag, Basel, Boston, Berlin 1997, pp.14-41.

Boole, George (1854), *The Laws of Thought*, Dover Publications.

Boole, George (later than 1854), 'Theory of Formal Logic', in *Selected Manuscripts on Logic and its Philosophy*, edited by Ivor Grattan-Guinness and Gérard Bornet, Birkhäuser Verlag, Basel, Boston, Berlin 1997, pp.112-116.

Boole, George (1855), 'Letter to Penrose', in *Selected Manuscripts on Logic and its Philosophy*, edited by Ivor Grattan-Guinness and Gérard Bornet, Birkhäuser Verlag, Basel, Boston, Berlin 1997, pp.200-202.

Boole, George (1856), 'On the Foundations of the Mathematical Theory of Logic', in *Selected Manuscripts on Logic and its Philosophy*, edited by Ivor Grattan-Guinness and Gérard Bornet, Birkhäuser Verlag, Basel, Boston, Berlin 1997, pp. 63-104.

Boole, George (1859), *A Treatise on Differential Equations*, Chelsea Publishing Company New York 1959.

Boole, George (1860a), *A Treatise on The Calculus of Finite Differences*, Chelsea Publishing Company New York 1958.

Boole, George (1860b), 'Logic', in *Selected Manuscripts on Logic and its Philosophy*, edited by Ivor Grattan-Guinness and Gérard Bornet, Birkhäuser Verlag, Basel, Boston, Berlin 1997, pp.126-153.

Boolos, George (1998), *Logic, Logic, and Logic*, Harvard University Press.

Boyer, Carl. B. (1989), *A History of Mathematics*, second edition, New York.

Brady, Geraldine (1997), 'From the Algebra of Relations to the Logic of Quantifiers', in *Studies in the Logic of Charles Sanders Peirce*, Indiana University Press, pp. 173-192.

References

Buridan, John (c. 1496/1500), *Sophisms On Meaning and Truth*, translated and with an introduction by Theodore kermit Scott, Appleton-century-Crofts, New York 1966.

Burris, Stanley. N. (1998) *Logic for Mathematics and Computer Science*, Prentice Hall.

Carnap, Rudolf (1934), *The Logical syntax of Language*, Routledge & Kegan Paul LTD.

Carroll, Lewis (1886), *Symbolic Logic & Game of Logic*, Mathematical Recreations of Lewis Carroll, Dover Publishing Company 1958.

Chang, C. C, and Keisler, H. J. (1973), *Model Theory*, North-Holland Publishing Company.

Church, Alonzo (1956), *Introduction to Mathematical Logic*, vol. 1, Princeton University Press.

Corcoran, J. and Wood, S. (1980), 'Boole's Criteria of Validity and Invalidity', in *Notre Dame Journal of Formal Logic*, 21, pp. 609-638.

Couturat, Louis (1985), *La Logique de Leibniz*, New York.

Currie, Gregory (1982), *Frege: An Introduction to his Philosophy*, Totowa, N. J. ; Barnes and Noble.

de Morgan, Augustus (1847), *Formal Logic*, London The Open Court Company 1926.

de Morgan, Augustus (1850), 'On the Symbols of Logic, the Theory of the Syllogism, and in Particular of the Copula', in *Transactions of the Cambridge Philosophical Society* 9, Part 1, 1851, pp.79-127.

de Morgan, Augustus (1860), 'On the Syllogism: IV; and on the Logic of Relations, in *Transactions of the Cambridge Philosophical Society* 10, pp.331-358. Repr. in *On the Syllogism and other Logical Writings*, ed. Peter Heath, Routledge & Kegan 1966.

Descartes, René (1955), *The Philosophical Works of Descartes*, Elisabeth S. Haldam & C. R. T. Ross, Dover Publication, vol. 1.

Dudman, H. V. (1976), 'From Boole to Frege' in *Studies on Frege I Logic and Philosophy of Mathematics*, Matthias Schirn, *Problemata Frommann-Holzbrog* 42, pp. 109-138.

Dummett, Michael (1981a), *The Interpretation of Frege's Philosophy*, Duckworth.

Dummett, Michael (1981b), *Frege: Philosophy of Language*, Harvard University Press, second edition.

Dummett, Michael (1991a), *The Logical Basis of Metaphysics*, Duckworth, London

Dummett, Michael (1991b), *Frege: Philosophy of Mathematics*, Duckworth, London

Dummett, Michael (1978), *Truth and Other Enigmas*, Duckworth.

Feys, Robert (1956), 'Boole as a Logician and Boole's Methods of Development and Interpretation', in *Transactions of The Royal Irish Academy* 57, pp. 97-112.

Frege, Gottlob (1879), *Begriffsschrift*, in *Conceptual Notation and Related Articles*, Terrell Ward Bynum, Oxford, Clarendon Press 1972, pp.101-208.

Frege, Gottlob (1879/91), 'Logic', in *Posthumous Writings*, Oxford UK: Basil Blackwell 1979, pp. 1-8.

Frege, Gottlob (1880/81), 'Boole's Logical Calculus and the Concept-script', in *Posthumous Writings*, Oxford UK: Basil Blackwell 1979, pp. 9-46.

Frege, Gottlob (1882a), 'Letter to Marty', in *Philosophical and Mathematical Correspondence*, Basil Blackwell Oxford 1980, p. 101.

Frege, Gottlob (1882b), 'On the Scientific Justification of a Conceptual Notation', in *Conceptual Notation and Related Articles*, Terrell Ward Bynum, Oxford, Clarendon Press 1972, pp. 83-89.

Frege, Gottlob (1882c), ' On the Aim of "Conceptual Notation"', in *Conceptual Notation and Related Articles*, Terrell Ward Bynum, Oxford, Clarendon Press 1972, pp. 90-100.

Frege, Gottlob (1882d), 'Boole's Logical Formula-language and my Concept-script', in *Posthumous Writings*, Oxford UK: Basil Blackwell 1979, pp. 47-52.

Frege, Gottlob (1884), *The Foundations of Arithmetic*, Northwestern University Press 1994.

Frege, Gottlob (1885), 'On Formal Theories of Arithmetic', in *G. Frege: Collected Papers on Mathematics, Logic, and Philosophy* edited by Brian McGuinness 1984, Basil Blackwell, pp. 112-121.

Frege, Gottlob (before 1891), 'Letter to Peano', in *Philosophical and Mathematical Correspondence*, Basil Blackwell Oxford 1980, p. 108.

Frege, Gottlob (1891), 'Function and Concept' in *Translations From the Philosophical Writings of G. Frege*, edited by Peter Geach & Max Black, Basil Blackwell Oxford 1970, pp. 21-41.

Frege, Gottlob (1892a), 'On Sense and Reference', in *Translations From the Philosophical Writings of G. Frege*, edited by Peter Geach & Max Black, Basil Blackwell Oxford 1970, pp 56-78.

Frege, Gottlob (1892b), 'On Concept and Object', in *Translations From the Philosophical Writings of G. Frege*, edited by Peter Geach & Max Black, Basil Blackwell Oxford 1970, pp 42-55.

Frege, Gottlob (1893), *The Basics Laws of Arithmetic*, University of California Press 1964.

Frege, Gottlob (1895a), 'A Critical Elucidation of some Points in E. Schröder, *Vorlesungen über die Algebra der Logik*', in *G. Frege: Collected Papers on Mathematics, Logic, and Philosophy* edited by Brian McGuinness 1984, pp. 210-228, Basil Blackwell.

Frege, Gottlob (1895b), 'Whole Numbers', in *G. Frege: Collected Papers on Mathematics, Logic, and Philosophy* edited by Brian McGuinness 1984, pp. 229-233, Basil Blackwell.

Frege, Gottlob (1895c), 'Letter to Hilbert', in *Philosophical and Mathematical Correspondence*, Basil Blackwell Oxford 1980, p. 33.

Frege, Gottlob (1897a), 'Logic', in *Posthumous Writings*, Oxford UK: Basil Blackwell 1979, pp. 126-151.

Frege, Gottlob (1897b), 'On Mr. Peano's Conceptual Notation and My Own', in *G. Frege: Collected Papers on Mathematics, Logic, and Philosophy*, edited by Brian McGuinness 1984, Basil Blackwell, pp. 234-248.

Frege, Gottlob (1898/99), 'Logical Defects in Mathematics', in *Posthumous Writings*, Oxford UK: Basil Blackwell 1979, pp. 157-166.

Frege, Gottlob (1899/1906), 'On Euclidean Geometry', in *Posthumous Writings*, Oxford UK: Basil Blackwell 1979, pp. 167-169.

Frege, Gottlob (1906), 'Introduction to Logic', in *Posthumous Writings*, Oxford UK: Basil Blackwell 1979, pp. 187-196.

Frege, Gottlob (1914), 'Logic in Mathematics', in *Posthumous Writings*, Oxford UK: Basil Blackwell 1979, pp. 204-252.

Frege, Gottlob (1918/9), 'Thoughts', in *G. Frege: Collected Papers on Mathematics, Logic, and Philosophy* edited by Brian McGuinness 1984, Basil Blackwell, pp. 351-372.

Frege, Gottlob (1919), 'Notes for Ludwig Darmstaedter', in *Posthumous Writings*, Oxford UK: Basil Blackwell 1979, pp. 253-257.

Frege, Gottlob (1924/25a), 'A New Attempt at a Foundation for Arithmetic', in *Posthumous Writings*, Oxford UK: Basil Blackwell 1979, pp. 278-281.

Frege, Gottlob (1924/25b), 'Sources of Knowledge of Mathematics and the Mathematical Natural Sciences', in *Posthumous Writings*, Oxford UK: Basil Blackwell 1979, pp. 267-274.

Gardner, Martin (1982), *Logic Machines and Diagrams*, University of Chicago Press, second edition.

Geach, P. T. (1962), *Reference and Generality: An Examination of Some Medieval and modern Theories*, Cornell University Press 1980.

Gillies, Donald (1992), *Revolutions in Mathematics*, edited by Donald Gillies, Oxford Science Publication, Clarendon Press.

Gillies, Donald (1993), *Philosophy of Science in The Twentieth Century*, Basil Blackwell.

Gödel, Kurt (1930), 'The Completeness of The Axioms of The Functional Calculus of Logic', in Van Heijenoort (1967), *From Frege to Gödel*, Harvard University Press, pp. 582-91.

Gödel, Kurt (1931), 'On Formally Undecidable Propositions of Principia Mathematica and Related Systems I', in Van Heijenoort (1967), *From Frege to Gödel*, Harvard University Press, pp. 596-616.

Gödel, Kurt (1951), 'Some Basic Theorems on The Foundations of Mathematics and Their Implication', in *Kurt Gödel: Collected Works*, vol. iii, edited by Solomom Feferman 1995, Oxford University Press, pp.304-323.

Goldfarb, Warren D. (1979), 'Logic in the Twenties: The Nature of the Quantifier', in *Journal of Symbolic Logic*, vol. 44, No. 3. (Sep., 1979), (accessible at http://uk.jstor.org/journals/00224812.html), pp. 351-368.

Grattan-Guinness, I. (1997), 'Boole's Quest for the Foundations of Logic', in *Selected Manuscripts on Logic and its Philosophy*, edited by Ivor Grattan-Guinness and Gérard Bornet, Birkhäuser Verlag, Basel, Boston, Berlin, 1997, pp. xiv-lxiv.

Hackett, F. E. (1955), 'The Method of George Boole', in *Proceedings of the Royal Irish Academy*, section A.57, pp. 79-87.

Hacking, Ian (1973), 'Leibniz and Descartes: Proof and Eternal Truths', in *Proceedings of the British Academy*, Oxford University Press, pp. 3-16.

Hailperin, Theodore (1976), *Boole's Logic and Probability*, North-Holland Publishing Company.

Hailperin, Theodore (1984), 'Boole's Abandoned Propositional Logic', in *History and Philosophy of Logic* 5 (1), p. 39-48.

Hamilton, W. (1852), *Discussion on Philosophy and Literature*, London.

Hamilton, William (1873), *Lectures on Metaphysics and Logic*, Boston, vol. 2.

Heijenoort, J. V. (1967a), *From Frege to Gödel, A Source Book in Mathematical Logic*, 1879-1931, Harvard University Press.

Heijenoort, J. V. (1967b), 'Logic as Calculus and Logic as Language', in *Synthese*, vol. 17,1967, pp. 324-330.

Heijenoort, J. V. (1970), *Frege and Gödel: Two Fundamental Texts in Mathematical Logic*, Cambridge Massachusetts.

Herbrand, Jacques (1930), 'Investigations in Proof Theory: The Properties of True Propositions', in Van Heijenoort (1967), *From Frege to Gödel*, Harvard University Press, pp. 529-581.

Hilbert, David (1904), 'On the Foundations of Logic and Arithmetic', in Van Heijenoort (1967), *From Frege to Gödel*, Harvard University Press, pp. 130-138.

References

Hilbert, David (1925), 'On the Infinite', in Van Heijenoort (1967), *From Frege to Gödel*, Harvard University Press, pp. 367-392.

Hilbert and Ackermann (1928), *Principles of Mathematical Logic*, Chelsea Publication Company 1950.

Houser, N.,Roberts, D. D & Evra, J. V. (1997) *Studies in the Logic of Charles Sanders Peirce*, Indiana University Press.

Huntington, E. V. (1904), 'Sets of Independent Postulates for the Algebra of Logic', in *Transactions of the American Mathematical Society*, vol. 5 and vol.35, pp. 288-309.

Huntington, E. V. (1933), 'Sets of Independent Postulates for the Algebra of Logic', in *Transactions of the American Mathematical Society*, vol.35, pp. 274-304.

Jevons, W. Stanley (1864), *Pure Logic and Other Minor Works*, Burt Franklin, New York 1971.

Jevons, W. Stanley (1874), *The Principles of Science*, Dover Publication, New York 1958.

Kneale, William (1947), 'Boole and the Revival of Logic', in *Mind* n.s 57, pp. 149-175.

Kneale, W & M. (1962), *The Development of Logic*, Clarendon Press.

Kuhn, Thomas (1962), *The Structures of Scientific Revolutions*, University of Chicago Press, second edition 1970.

Laita, L. (1977), 'The Influence of Boole's Search for a Universal Method in Analysis on the Creation of his Logic', in *Annals of Science* 34, pp. 163-176.

Lakatos, Imre (1970), 'Falsification and the Methodology of Scientific Research Programmes', in *Criticism and The Growth of Knowledge*, edited by Imre Lakatos and Alan Musgrave Cambridge University Press, pp. 91-189.

Lakatos, Imre (1976), *Proofs and Refutations: The Logic of Mathematical Discovery*, edited by J Worrall and E G Zahar.

Lear, Jonathan (1980), *Aristotle and Logical Theory*, Cambridge University Press.

Lee, Harold Newton (1961), *Symbolic Logic*, Routledge & Kegan Paul Limited London.

Leibniz, G. W. (1765), *News Essays*, edited by Peter Remuant and Jonathan Bennett, Cambridge University Press 1996.

Levine, James (1986), 'Logic and Truth in Frege', in *Proceedings of Aristotelian Society*, Supplementary Volume 70, pp. 141-175.

Lewis & Langford (1959), *Symbolic Logic*, Dover Publication, INC.

Lewis, C. I. (1960), *A Survey of Symbolic Logic*, Dover Publication, New York.

Locke, John (1710), *An Essay Concerning Human Understanding*, sixth edition, London, H. Hills, vol. 1.

Löwenheim, Leopold (1915), 'On Possibilities in the Calculus of Relatives', in Van Heijenoort (1967), *From Frege to Gödel*, Harvard University Press, pp. 232-251.

Lukasiewicz, J. (1951), *Aristotle's Syllogistic From the Standing Point of Modern Formal Logic*, Oxford.

Macfarlane (1885), 'The Logical Spectrum', in *Philosophical Magazine*, vol. 19.

MacHale, D. (1985), *George Boole: His Life and Work*, Boole Press, Dublin.

Marquand, A. (1881), 'A Logical Diagram for n Terms', in *Philosophical Magazine*, vol. 12.

McColl, Hugh (1880), 'Symbolical Reasoning', in *Mind*, vol. 5, Issue 17 (Jan., 1880), (accessible at http://www.jstor.ac.uk/journals/00264423.html), pp. 45-60.

Merrill, D. D. (1997), 'Relations and Quantification in Peirce's Logic, 1870-1885', in *Studies in the Logic of Charles Sanders Peirce*, Indiana University Press, pp. 158-172.

Mill, J. S. (1884), *A System of Logic*, People's edition London.

Moody, E. A. (1953), *Truth and Consequence in Mediaeval Logic*, North-Holland Publishing Company, Amsterdam.

Noonan, W. Harold (2001), *Frege: A Critical Introduction*, Polity Press.

Peano, Guiseppe (1888), 'The Operation of Deductive Logic', in *Selected Works of Guiseppe Peano 1973*, edited by H.C. Kennedy, Allen & Unwin, pp. 75-100.

Peano, Guiseppe (1889), 'The Principles of Arithmetic, Presented by a New Method' (1889), in *Selected Works of Guiseppe Peano 1973*, edited by H.C. Kennedy, Allen & Unwin, pp. 101-134.

Peano, Guiseppe (1891), 'The Principles of Mathematical Logic', in *Selected Works of Guiseppe Peano 1973*, pp.153-161, edited by H.C. Kennedy, Allen & Unwin.

Peirce, C. S. (1865), 'Lecture VI: Boole's Calculus of Logic', in *Writings of Charles S. Peirce* vol. 1, Indiana University Press, Bloomington, pp. 223-239.

Peirce, C. S. (1867), 'On an Improvement in Boole's Calculus of Logic', in *Writings of Charles S. Peirce* vol. 2, Indiana University Press, Bloomington, pp. 12-23.

References

Peirce, C. S. (1870), 'Description of a Notation for the Logic of Relatives, Resulting from an Amplification of the Conceptions of Boole's Calculus of Logic', in *Writings of Charles S. Peirce* vol. 2, Indiana University Press, Bloomington, pp. 359-429.

Peirce. C. S. (1880), 'On The Algebra of Logic', in *American Journal of Mathematics*, vol. 3, No. 1, (accessible at http://uk.jstor.org/journals/00029327.html), pp. 15-57.

Peirce, C. S. (1882a), 'On the Logic of Relatives', in *Writings of Charles S. Peirce* vol.4, Indiana University Press, Bloomington, pp. 336-341.

Peirce, C. S. (1882b), 'Letter, Peirce to O. H. Mitchell', in *Writings of Charles S. Peirce* vol. 4, Indiana University Press, Bloomington, pp. 394-399.

Peirce, C. S. (1884), 'On the Algebra of Logic', in *Writings of Charles S. Peirce* vol. 5, Indiana University Press, Bloomington, pp.107-115.

Peirce, C. S. (1885a), 'On the Algebra of Logic: A Contribution to the Philosophy of Notation', in *Writings of Charles S. Peirce* vol. 5, Indiana University Press, Bloomington, pp.162-190.

Peirce, C. S. (1885b), 'Notes on the Algebra of Logic', in *Writings of Charles S. Peirce* vol. 5, Indiana University Press, Bloomington, pp.191-203.

Peirce, C. S. (1885c), 'Studies in Logical Algebra', in *Writings of Charles S. Peirce* vol. 5, Indiana University Press, Bloomington, pp. 204-220.

Peirce. C. S.(1893), 'The Aristotelian Syllogistics', in *Collected Papers of C. S. Peirce*, edited by Charles Hartshorne and Paul Weiss Cambridge, Harvard University Press, vol. ii, pp. 279-283.

Popper, Karl (1970), 'Normal Science and its Danger', in *Criticism and The Growth of Knowledge*, edited by Imre Lakatos and Alan Musgrave Cambridge University Press, pp. 51-58.

Popper, Karl (1974), 'Replies to My Critics', in *The Philosophy of Karl Popper*, edited by Paul Arthur Schilpp, Northwestern University, vol. 2, pp. 961-1174.

Putnam, Hilary (1992), *Realism With a Human Face*, Harvard University Press.

Quine, W. V. O. (1961), *Mathematical Logic*, Cambridge Harvard University Press.

Quine, W. V. O. (1966), *Selected Logic Papers*, Random House New York.

Quine, W. V. O. (1970), *Philosophy of Logic*, Prentice, INC.

Quine, W. V. O. (1974), *Methods of Logic*, 3rd edition Routledge & Kegan.

Quine, W. V. O. (1987), *Quiddities*, The Belknap Press of Harvard university Press.

Ricketts, T. (1986), 'Logic and Truth in Frege', in *Proceedings of Aristotelian Society*, Supplementary Volume 70, pp. 121-140.

Russell, Bertrand (1900), *A Critical Exposition of The Philosophy of Leibniz*, Allen & Unwin House- London.

Russell, Bertrand (1903), *The Principles of Mathematics*, London, second edition 1937.

Russell, Bertrand (1919), *Introduction to Mathematical Philosophy*, Routledge 1993.

Russell, Bertrand (1922), *Our Knowledge of The External World*, George Allen & Unwin House- London.

Russell, B, Whitehead, Alfred (1927), *Principia Mathematica*, vol. 1 Cambridge University Press, second edition.

Scholz, H. (1931), *Esquisse d'une histoire de la logique*, Paris.

Schröder, E. (1891-95), *Vorlesungen über die Algebra der Logik*, vol. ii and iii Chelsea Publishing Company Bronx, New York 1966.

Sheffer, H. M. (1913), 'A Set of Five Independent Postulates for Boolean Algebras', in *Transactions of the American Mathematical Society*, vol.14, pp. 481-488.

Skolem, T. (1920), 'Logico-Combinatorial Investigations', in Van Heijenoort (1967), *From Frege to Gödel*, Harvard University Press, pp. 254-263.

Skolem, T. (1922), 'Some Remarks on Axiomatized Set Theory', in Van Heijenoort (1967), *From Frege to Gödel*, Harvard University Press, pp. 291-301.

Sluga, Hans (1980), *Frege*, Routledge & Kegan Paul.

Sluga, Hans (1987), 'Frege Against the Booleans', in *Notre Dame Journal of Formal Logic*, vol. 28, 1, pp. 80-98.

Smith, G. C. (1983), 'Boole's Annotations on the Mathematical Analysis of Logic', in *History and Philosophy of Logic* 4 (1) pp. 27-38.

Stanley, W. (1919), *Principle of Logic*, London and Edinburgh.

Stern, A. (1988), *Matrix Logic*, North-Holland Amsterdam.

Styazhkin, N. I. (1969), *History of Mathematical Logic From Leibniz to Peano*, The M. I. T. Press.

Stebbing, L. S. (1950), *A Modern Introduction to Logic*, Methuen & Co. LTD. London.

Tarski, Alfred (1956), *Logic, Semantics, Metamathematics*, Hackett Publishing Company 1983.

Tarski, Alfred (1941), *Introduction to Logic and to the Methodology of Deductive Sciences*, Oxford University Press, New York.

Van Evra, J. (1977), 'A Reassessment of George Boole's Theory of Logic', in *Notre Dame Journal of Formal Logic* 18, pp. 363-377.

References

Venn, John (1876), 'Boole's Logical System', in *Mind*, vol. 1, No. 4. (Oct., 1876), (accessible at http://uk.jstor.org/journals/00264423.html), pp. 479-491.

Venn, John (1881), *Symbolic Logic*, Chelsea Publishing Company 1971.

Weiner, Joan (1990), *Frege in Perspective*, Cornell University Press.

Weiner, Joan (1999), *Frege*, Oxford University Press.

Whateley, Richard (1826), *Elements of Logic*, New York, 1925.

Whitehead, Alfred, N. (1898), *A Treatise on Universal Algebra*, vol. 1. Cambridge University Press.

Wilder, Raymond (1965), *Introduction to The Foundations of Mathematics*, John Wiley & Sons, INC.

Wittgenstein, Ludwig (1914), *Notebooks*, edited by G. H. Von Wright and G. E. M Anscombe, Basil Blackwell Oxford 1969.

Wittgenstein, Ludwig (1913), *Tractacus Logico-Philosophicus*, Routledge & Kegan Paul, London 1974.

Wright, Crispin (1983), *Frege's Conception of Numbers as Objects*, Aberdeen University Press.

www.ingramcontent.com/pod-product-compliance
Lightning Source LLC
Chambersburg PA
CBHW070556100426
42744CB00006B/299